THE BIG CATS AND THEIR FOSSIL RELATIVES

THE BIG CATS
AND THEIR FOSSIL RELATIVES

An Illustrated Guide to Their Evolution and Natural History

Illustrations by
MAURICIO ANTÓN

Text by
ALAN TURNER

COLUMBIA UNIVERSITY PRESS
New York

Columbia University Press
Publishers Since 1893
New York   Chichester, West Sussex
Copyright © 1997 Columbia University Press
All rights reserved

Library of Congress Cataloging-in-Publication Data
Turner, Alan, 1947–
      The big cats and their fossil relatives : an illustrated guide to their evolution
   and natural history / text by Alan Turner ; illustrations by Mauricio Antón.
         p. cm.
   Includes bibliographical references (p. 221) and index.
   ISBN 978-0-231-10228-5 (cloth) — ISBN 978–0–231–10229–2 (pbk.)
   1. Felidae. 2. Felidae, Fossil. 3. Felidae—Evolution. 4. Felidae—Pictorial
   works. 5. Felidae, Fossil—Pictorial works. I. Antón, Mauricio. II. Title.
   QL737.C23T83   1996
   599.74'428—dc20   96-3969

♾

Casebound editions of Columbia University Press books are printed on
permanent and durable acid-free paper.
Printed in U.S.A.
c 10 9 8 7 6 5 4 3 2
p 10 9 8 7

Frontispiece: A group portrait of living large cats
All living large cats can be distinguished at first sight, although their skeletons
differ less from each other than they do from any of the specialized felid saber-
tooths. The jaguar (bottom, center) differs from the leopard in its stockier build
and in the pattern of larger, closed rosettes on its coat. The clouded leopard
(resting on the branch) and the snow leopard (below it) both have unmistakable
coats that differ from that of the true leopard. The male lion (center) has a
unique appearance, but even the maneless females can be distinguished from a
puma (center, on trunk) by their large, squarish heads on high shoulders, their
long legs, and their black-tufted tails.

For our families
and in memory of Björn Kurtén

CONTENTS

FOREWORD

ONLY SEVEN EXTANT LARGE-BODIED CATS SURVIVE FROM THE important and remarkably diverse radiations of felid carnivores since mid-Tertiary times, some 25 million years (Ma) ago. Fortunately these surviving felines are quite readily distinguished morphologically, differ substantially in body size, exhibit some distinctive predatory behaviors and killing practices, and manifest a variety of territorial behaviors and social groupings. Thus the extant species of the genera *Panthera* (lion, leopard, tiger, jaguar, and snow leopard), *Felis* (puma or cougar), and *Acinonyx* (cheetah) afford invaluable insights into the adaptations, diversifications, and behaviors of one tribe (Felini) of the ever-fascinating, however fearsome, big cats. There is a fundamental human attraction, tempered by trepidation, toward these animals, among the most visited and admired in zoological gardens and much prized when observed in natural habitats. Yet all have shrunk substantially in geographic distribution within recorded history, and they have experienced significant reduction in numbers in this century.

This scant representation of extant species is but a shadow of the greater felid diversity, broader geographic distribution, and (probably) enhanced local and overall population numbers reflected by the fossil record. Extinctions in the course of the late Tertiary and at several times during the Pleistocene led to the loss of the last of the paleofelids (*Barbourofelis*) and of three other major groups (tribes) of felids—metailurines, homotherines, and smilodontines—comprising minimally 8 genera and, very conservatively, 24 species. In addition, at least 5 larger species of the tribe Felini experienced Pleistocene extinction. Thus, minimally, about four-fifths of the large-bodied felids known over the past 10–12 Ma have vanished—as, it should be noted, have innumerable ungulate and other species on which they preyed over five continents. This is a vast loss, a consequence of natural processes attending upon extinction(s), and a phenomenon especially worthy of consideration in these times of preoccupation with biodiversity and its conservation. The causes of such disappearances are of central concern to paleobiologists, although there is limited agreement as to the roles played by the effects of global climatic change and its attendant habitat fragmentation, transformation, and replacement; by interspecific competition; or by turnovers and other compositional and structural changes in prey communities. These and other factors have been commonly invoked to account for extinctions, but the actual processes and circumstances attendant thereon are, more often than not, matters of inference and speculation.

This remarkable book is an extensively illustrated exposition of the origins, diversifications, and adaptations of the larger-bodied members, extant and extinct, of the cat family, the Felidae (it thus excludes the clouded leopard, marbled cat, and lynxes). The structure, adaptations, and behaviors of the seven extant species constitute an appropriate and invaluable perspective from which other, often quite different genera and their included species may be compared and evaluated, and their adaptations and associated behaviors thereby elucidated. The vagaries of the fossil record are such that past animals are often known only quite incompletely and imperfectly, and frequently only from partial fragments of skull, jaw, and/or dental elements. This is particularly the case among those kinds that are relatively few in overall numbers, occur naturally in small groups or lead solitary lives, and for whatever reason have lesser chances overall to be preserved after death. Fortunately, their distinctiveness can be recognized from characters commonly expressed in skull and tooth anatomy, even when the total animal is initially largely or wholly unknown skeletally. This has been true of course for large-bodied felids, almost all of which were initially known only in fragmentary states. Ultimately, more complete remains tend to come to light, in time including even entire skeletons of one or more individuals. Such has been the history of the knowledge and understanding of the extinct saber-toothed genera *Homotherium* and *Megantereon* (a complete skeleton of each recovered from the ancient volcanic maar of Senèze in the French Auvergne, and American species of the former from Friesenhahn Cave in Texas), and *Machairodus* (largely complete skeletons from Cerro Batallones in central Spain and from several midwestern American localities). Now the long ill-known, even enigmatic *Paramachairodus* is known from multiple skeletons (again from Cerro Batallones), and *Dinofelis*, first recorded from eastern Asia, is now well known skeletally from karstic cavern infills of the Transvaal, South Africa. The vast number of individual remains of a *Smilodon* species, California's state fossil, represented at the Rancho La Brea tar seeps in the Los Angeles basin, are altogether unique and otherwise unparalleled owing to the special sedimentary circumstances of such natural traps.

Alan Turner's lucid text affords a fine introduction to these bigger felids—their place in nature, occurrence and preservation in the fossil record, profiles of their individual (taxonomic) identities, and manifold details of their principal anatomical, physiological, social, and behavioral characteristics in relation to life ways, subsistence, and associated adaptations. The place of extinct forms in past faunal associations and in guilds (a term referring to a group of animals that exploit a similar resource), their eventual disappearances, and the appearance and dispersals of modern representatives are thoroughly elucidated. The fine drawings and paintings of Mauricio Antón greatly facilitate and visually enhance the textual exposition; their individual legends afford much additional information and the requisite elaboration of particular subject matters. Antón's skill and expertise in animal anatomy are evidenced throughout the volume, particularly in the exceptional reconstructions that capture both extinct and extant forms in natural settings, often in full-color plates. These insightful renditions are absolutely outstanding and fully testify to the artist's unparalleled knowledge and his craftsmanship at bringing the past to life.

This volume offers many valuable insights into the evolution and natural history of the greater cats, both present and past. The authors' knowledge of the subject is intimate and profound, and the talents of the scientific illustrator offer elegant visual counterpoint of exceptional merit. The book—appropriately dedicated to the late and lamented Björn Kurtén, an outstanding contributor to studies of carnivore evolution and paleobiology—fully deserves and will surely receive a wide readership.

F. Clark Howell
Laboratory for Human Evolutionary Studies
University of California, Berkeley

PREFACE

FOSSILS ARE THE RECORD OF THE DEVELOPMENT OF LIFE ON earth, and their existence is widely if imprecisely known. If you ask people at random to give an example of a fossil, at least one of the following three is almost certain to be mentioned: dinosaurs, mammoths, and saber-toothed cats have become fixed in the public mind, although usually in a confused jumble that assumes all were living at the same time as our own prehistoric, cave-dwelling ancestors. In fact, the dinosaurs have been extinct for 65 million years (65 Ma), gone long before cats, mammoths, or humans appeared on the scene.

Fossil cats have been known to paleontologists ever since the systematic collecting of skeletal remains began, and a vast number of scientific papers have now been written on the subject. Descriptions of specimens were published in France in the first half of the nineteenth century, including observations on the canine teeth of what is now known to be the saber-toothed cat genus *Homotherium* (although the eminent French paleontologist Georges Cuvier considered that they belonged to a bear). The genus *Smilodon* was given its name by the Danish paleontologist Peter Wilhelm Lund as long ago as 1842, on the basis of bones that he had found in Brazilian caves of upper Pleistocene age. Lund was so impressed by the enormous canines of the animal that he gave it the species name *populator*, meaning "he who brings devastation." In 1867 the British paleontologists William Boyd Dawkins and William Sanford published a richly illustrated catalogue listing the specimens of felids (mainly lions) found in the caves of the Mendip region to the south of Bristol, and between 1866 and 1872 they published an extensive work on the British Pleistocene Felidae (again, mainly lions). Such was the number of different species known toward the end of the nineteenth century that as long ago as 1880 the American paleontologist Edward Cope attempted to bring order to the study by offering a detailed appraisal of numerous names and their likely relationships.

But while numerous popular volumes have been written on the subject of dinosaurs, and the mammoths have received at least their share of publicity, little has been said in detail about the cats, and about their evolution and fossil relatives, in a way that is available to the nonspecialist. In this book we therefore aim to provide an authoritative and yet popular and accessible account of the evolution of the larger cats by bringing together the evidence of modern behavior and the fossil record. Throughout, we refer to both living and extinct taxa in the discussions, since the study of both is necessary for an integrated understanding of the evolution and natural history of these fascinating animals. The modern cats show us how such animals may live and act, while the fossils show us the vast range of species that have existed with a broadly feline morphology and the broader pattern of change over time.

We have restricted our detailed attention to the larger species: they are best represented in the fossil record, and it is the larger animals among the living representatives that hold the greatest fascination for many people. This decision is in some senses arbitrary, not the least because of the wide range of variation in size and weight found among members of the living species. Our criterion takes a lowermost body weight of around 40 kg, and it thereby includes cats such as the America puma and the Afro-Asian leopard but excludes others such as the lynxes and the clouded leopard of southeastern Asia. It may be objected that lynxes and clouded leopards are still quite large animals—but if we include them then one might reasonably ask, why not continue down to the smallest of species; the list then grows, and the book expands beyond reasonable size.

All this is not to say that we exclude the smaller cats from any consideration, merely that we give emphasis to their larger relatives. Much can be learned about cats from observing the common domestic animal. Structurally, it can be seen as simply a scaled-down model of a lion or a leopard, and in evolutionary terms the larger cats may even be considered as scaled-up versions of something much like a domestic cat. The livingroom pet has the same overall body plan, with long limbs, claws, and sharp teeth well suited to catching, dispatching, and eating its prey. It behaves, hunts, and deals with its food in a thoroughly catlike manner, and is therefore as much a member of the family as is the lion, the leopard, or one of the extinct saber-tooths. It is also a more convenient example to refer to in the comfort of the home when some detail of anatomy or behavior is mentioned.

What we present is the result of a collaboration between an artist interested in evolution, natural history, and the reconstruction of fossil animals, and a paleontologist interested in seeing what the dry bones in the laboratory might have looked like with flesh and fur on them. The illustrations are therefore an integral part of the discussion, and have for the most part been specially made to accompany the text; those of the fossil species, in particular, have been based directly on the skeletal evidence available and are not merely slightly altered versions of living cats with the addition of large fangs. We hope that the result will bring these fascinating animals to life.

The plan of the book reflects our intention to interweave the evidence from fossil and living species. We begin in chapter 1 with an overview of the place of cats in nature and how we name them, and a brief outline of what fossils actually are and how we find them and work out their ages. In chapter 2 we explain some of the principles of evolution and summarize the earliest history of the cats from their first appearance in the fossil record some 30 Ma ago. Chapter 3 then provides more detailed information about the individual species that we know from the fossil record and that we find living today in various parts of the world, including a summary of the fossil history of each living species. In chapter 4 we describe how cats work. This chapter gives close attention to why their eyesight is so good, how they use their claws and teeth, and how they move, and it ends by showing how we can use our knowledge about living animals to reconstruct not only the anatomy but also the movement and thus aspects of the behavior of extinct cats. Chapter 5 then integrates the evidence of anatomy and movement with observations of social and hunting activity, again showing how we can infer much about the extinct species when we put various lines of evidence together. Chapter 6 takes a broad, long-term look at the changes that have occurred on the earth over the past several millions of years, and examines the evolution of the larger cats within this wider framework. The chapter ends with a list of some of the places where fossil cats may be seen and is followed by some suggestions for further reading for those who wish to have more information about any aspect of our discussion.

ACKNOWLEDGMENTS

THIS WORK WOULD HAVE BEEN IMPOSSIBLE WITHOUT ACCESS TO collections held in institutions in Europe, Africa, and North America and the opportunity to study, measure, and photograph the fossil and modern specimens at first hand. We thank all those responsible for such collections, in particular

Jordi Agustí and Salvador Moya-Sola, Institut Paleontológic Miquel Crusafont (Sabadell); • Luis Alcalá, Museo Nacional de Ciencias Naturales (Madrid); • Jesús Alonso, Museo de Ciencias Naturales de Alava (Vitoria); • Margarita Belinchón, Museu Paleontológic Municipal de Valencia (Valencia); • Angel Galobart, Museo Arquaeológic Comarcal de Banyoles (Banyoles); • Judy Maquire, Bernard Price Institute for Palaeontological Research (Johannesburg); Léonard Ginsburg, Germaine Petter, and Françoise Renoult, Muséum National d'Histoire Naturelle (Paris); • Anthony Stuart, Castle Museum (Norwich); • Elmar Heitzmann, Staatliches Museum für Naturkunde (Stuttgart); • Michel Philippe, Musée Guimet d'Histoire Naturelle (Lyon); • Abel Prier and Rolland Ballesio, Université Claude Bernard (Lyon); • Marie-Françoise Bonifay, Laboratoire du Quaternaire, CNRS Luminy (Marseille); • Inesa Vislobokova, Palaeontological Institute of the Russian Academy of Sciences (Moscow); • Marina Sotnikova, Geological Institute of the Russian Academy of Sciences (Moscow); • Lorenzo Rook and Frederico Masini, Universitá degli Studi (Florence); • Peter Andrews, Andrew Currant, Jerry Hooker, and Julliet Clutton-Brock, The Natural History Museum (London); • Richard Tedford, American Museum of Natural History (New York); • Brett Hendey, South African Museum (Cape Town); • David Wolhuter, Francis Thackeray, and David Panagos, Transvaal Museum (Pretoria); • Hans-Dietrich Kalhke, Ralf-Dietrich Kahlke, and Lutz Maul, Institut für Quartärpaläontologie (Weimar); • Jens Franzen, Forschungsinstitut Senckenberg (Frankfurt); • and Adrian Friday, Zoological Museum (Cambridge).

We owe a great debt to many friends and colleagues for discussion, advice, and encouragement, especially Dan Adams, Emiliano Aguirre, Jordi Agustí, Luis Alcalá, Peter Andrews, Rolland Ballesio, Jon Baskin, Gérard de Beaumont, Andrew Currant, Giovanni Ficcarelli, Ann Forsten, Mikael Fortelius, Rosa García, Léonard Ginsburg, Francisco Goin, Brett Hendey, Clark Howell, Ralf-Dietrich Kahlke, the late Björn Kurtén, Adrian Lister, Martin Lockley, Gregori López, Larry Martin, Jay Matternes, Plinio Montoya, Jorge Morales, Manuel Nieto, Germaine Petter, Robert Santamaría, José Luis Sanz, Chris Shaw, Andrei Sher, Dolores Soria, Marina Sotnikova, Fred Spoor, Richard Tedford, Danilo Torre, Blaire Van Valkenburg, Lars Werdelin, Bernard Wood, and Xiaoming Wang. We are particularly grateful to Harold Bryant for providing us with material on nimravid phylogeny in advance of publication.

It is also our pleasure to thank Clark Howell for agreeing to write a foreword to the book. His own work on fossil carnivores, often undertaken in collaboration with Germaine Petter, has been a source of information and inspiration to us.

CREDITS

The painting in plate 14 is exhibited at the Institut Paleontológic Miquel Crusafont, Sabadell, Spain.

The painting in plate 16 was first shown in the exhibition "Madrid antes del Hombre" organized by the Comunidad Autónoma de Madrid and the Museo de Ciencias Naturales, Madrid, Spain.

THE BIG CATS AND THEIR FOSSIL RELATIVES

CHAPTER 1
Cats: Their Place in Nature

THE EARTH IS ROUGHLY 4,500 MILLION YEARS OLD. FOR THE PAST 250 Ma at least there have been mammals living and dying on it. We ourselves, along with the cats, are members of the class Mammalia, a class defined by, among other things, the possession of hair, mammary glands, and the birth of live young. As such, we belong in one of the three-fold divisions of the class, the placental mammals; the other two divisions consist of pouched marsupials (such as kangaroos) and egg-laying mono-tremes (such as the duck-billed platypus). Together with the other animals that have a bony skeleton, such as the dinosaurs, the birds, or the fishes, the mammals make up the phylum Chordata, often known as the vertebrates. That bony skeleton has proved to be the source of an excellent fossil record for the vertebrates.

A detailed account of the skeletal and muscular anatomy of the cats will be given in chapter 4, but since the teeth and the skeleton will form such an important part of our discussions it is important that we introduce some of the terminology to be used. Figure 1.1 shows the skeletons of two typical cats—one living, the lion, and one extinct, the American saber-toothed species known as *Smilodon fatalis*; while figure 1.2 shows the skull and teeth of a living leopard. As you can see, despite differences in bodily proportions and in details of the skull and dentition, the overall similarities between the lion and *Smilodon* are great. Both animals have the same bones in the same place. Basically, so do all vertebrates, although some considerable modifi-cations have taken place in some lineages, with the loss of limbs (as in whales) or their extreme modification (as in seals). It is this consistency that allows us to name the bodily parts. We can refer to the upper arm bone as the humerus, whether we deal with a living cat, an extinct saber-tooth, or even ourselves.

Around 65 Ma ago the dinosaurs went extinct and the mammals became the dominant class of larger terrestrial animals. At that stage few of the species bore any close resemblance to those that exist today, although broad similarities may be observed in the fossil remains. Since that time species have come and gone; whole families of related animals have become extinct, and new ones, such as our own, have arisen. The cats—the Felidae, to give them their proper designation—are one such family.

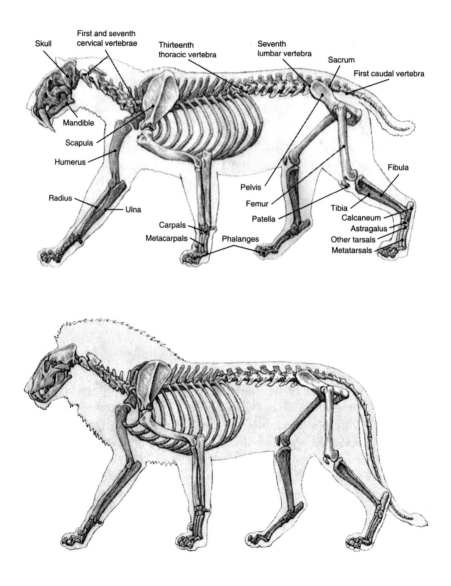

FIGURE 1.1 *Skeletons of a saber-toothed cat,* Smilodon fatalis *(top), and a lion,* Panthera leo The figure shows the essential similarities between the skeletons of two very distantly related cats and provides an introduction to the names of the various bones.

NOMENCLATURE

Why do we refer to the cats as a family? In order to make sure we all refer to the same thing each time we talk of a given type of animal, a formal system of naming has been established, based on principles laid down more than two hundred years ago by the Swedish naturalist Carl Linnaeus. What we recognize as a particular kind of animal, be it a sheep or a wolf, is known

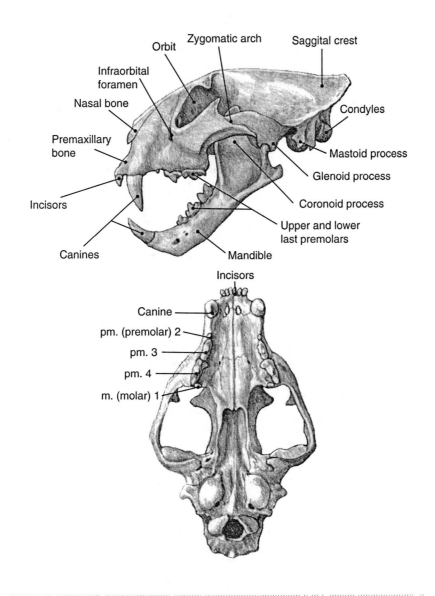

FIGURE 1.2 *Skull of a leopard,* Panthera pardus, *showing the nomenclature of the teeth and major features.*

as a species. A species has a formal, two-part name in Latin, consisting of the specific name itself and the genus. A genus is a group of species considered to be more closely related to each other than to any other species; in other words, it is considered that they share a common ancestral species. The genus name begins with a capital letter, and the whole name is usually written in italics.

Thus the lion has the species name *Panthera leo*, which consists of the specific name *leo* in the genus *Panthera*, and is distinguished by that name from the leopard, *Panthera pardus*, and from the tiger, *Panthera tigris*. These three members of the genus are thought to be more closely related to each other than they are to the cheetah, for example, which is placed in a separate genus and formally known as *Acinonyx jubatus*. (In a sentence or paragraph where the same genus name is used several times it is often shortened to just its initial letter after the first usage; thus *Panthera leo*, *P. pardus*, and *P. tigris*—unless there is any possibility of confusion.) Of course many species have common names, such as the lion, the leopard, or the tiger, but many fossil species simply do not, and we then have no choice but to use the scientific term. To the nonspecialist it may seem a clumsy way of naming an animal, but it does ensure precision and it remains the same whatever the language of the person speaking or writing about the animal. *Smilodon fatalis* can be unambiguously identified only by that name. Nearly everyone will be familiar with at least one formal name: the dinosaur *Tyrannosaurus rex* is famous the world over, and the fact that it is known only by its scientific name has proved no barrier to public interest or cinematic fame.

The various genera are in turn placed in a family, and thus we have the family Felidae for the cats. Felidae is the formal name; informally, they are often referred to as felids. The various dog genera are similarly placed in the family Canidae (canids), and the hyenas in the family Hyaenidae (hyaenids); and all three, together with other families such as the Ursidae (the bears), are in turn placed within the order Carnivora.

The term "Carnivora," together with its informal version "the carnivores," should be clearly understood, however. Both refer to a group (the order) of animals considered to be closely related. But the word "carnivore" also has the everyday meaning of an animal that eats meat. We eat meat, and are therefore carnivorous, but we are not members of the Carnivora: we are placed in the order Primates, along with monkeys and apes (some of whom may also eat meat on occasion). More confusingly still, some members of the Carnivora have evolved in a particular way and no longer eat just meat; the giant panda, with its well-known love of bamboo, is perhaps the best-known and most frequently quoted example, although it will take meat and retains a digestive system fully capable of handling such food. But it is important that this difference of terminology is observed, and throughout this volume the term "carnivore" should be understood to refer to formal membership in the Carnivora, rather than simply the eating of meat—unless specifically stated otherwise, or the context makes the point abundantly clear.

Together with other various orders—such as the Rodentia (the rodents), the Artiodactyla (the even-toed ungulates with split hooves, such as antelopes or deer), and our own order, the Primates—the Carnivora belong to the class Mammalia. This ranked system of naming, applied to the plant and the animal kingdoms (a higher category still) reflects opinions about the closeness of the relationships between the various species, genera, and so forth, and is useful to the zoologist, botanist, or paleontologist as a form of shorthand. Thus we have:

Kingdom: Animalia
Phylum: Chordata (Vertebrates)
Class: Mammalia
Order: Carnivora
Family: Felidae
Genus: *Panthera*
Species: *Panthera leo*

Of course this arrangement presents a very simplified picture of the ranking of just one species. We have already said that species considered to be most closely related to each other and descended from a common ancestral species are linked into the higher category of genus, and it is perhaps obvious that within a family some genera will be more closely related than others, and thus will form natural groups between the levels of family and genus. There are two such formal groupings, referred to as the subfamily and the tribe. But the allocation of genera to these groups is not always straightforward, since relationships are a matter of inference, and even the living cats have proved to be rather difficult to classify to the satisfaction of all. Including fossil taxa can complicate matters considerably. For the purpose of this volume we have elected to accept four tribes in two subfamilies to accommodate most of the specimens, both subfamilies being taken to derive from an earlier "stem" group containing the genera *Proailurus* and *Pseudaelurus*. Within this scheme the lion would be a member of the tribe Felini within the subfamily Felinae, or "true" cats with conical canines. The other subfamily of the Felidae is the Machairodontinae, the saber-toothed species with flattened and elongated upper canines, and they are usually placed in three tribes. We shall discuss this aspect of felid classification in more detail in the next two chapters.

FOSSILS: WHAT THEY ARE AND HOW THEY FORM

We have said that we are dealing here with larger living cats and their fossil relatives. But what are fossils, and how do they form?

It is sobering to realize that most of the animal species that have ever lived are now extinct. But many still occur as skeletons and teeth in the geological sediments that lie around us and beneath our feet. In some cases we even have evidence of the soft parts, perhaps the most extreme example being impressions of jellyfish: animals without any skeleton at all, but in which the body has been covered in sediments sufficiently fine to retain much of the detail. Those remains are fossils—a word that formerly applied to any object dug up but is now reserved for portions of once-living organisms preserved in the ground, although it may be extended to include traces of those organisms (in the form of footprints, for example). The latter, even in the absence of fossil bones, may give us considerable information, as we shall see in chapter 4.

The fossils we deal with are therefore the remains of animals that died in a variety of circumstances. The bony skeleton and teeth can survive extremely well in the right conditions, but we should bear in mind that the fossil record preserves only a small portion of the total number of animals of a species that lived in the past. This is because destructive processes will ensure that most skeletons do not survive. These processes include decay as the carcass lies on the ground, consumption by other animals, and perhaps destruction in the sediments that accumulate to form a geological stratum. Not every animal that dies, therefore, will become fossilized, and even bones that do survive to become fossils may be lost if the deposits are subsequently eroded.

Of all things, rapid burial is one of the most effective means of ensuring preservation. Living bone is a tissue consisting of organic material and a mineral matrix largely made up of calcium phosphate. While it may look inert, living bone bleeds if cut. When the animal dies the organic material of the skeleton, mostly the protein collagen, begins to decay, and in time the whole bone may simply disappear in sufficiently acidic conditions. But once covered, the bone may also become impregnated by ground waters that invade the matrix of the bone and replace part of it with dissolved minerals from the rocks and sediments above. This happens very typically when bones become buried in limestone caves, for example, as illustrated here in figure 1.3. Caves are by no means the only place to find fossils, but they are a common source in many parts of the world for the remains of larger cats that may have died during occupation or simply become trapped. One particularly good example of the latter is that of Natural Trap Cave in north-central Wyoming, excavated by Larry Martin, where complete skeletons of the American cheetah-like cat *Miracinonyx trumani* and remains of lions referred to the species *Panthera atrox* have been found associated with the skeletons of other carnivores and prey species. In many cases the deposit of such a cave will be an easily worked material often known informally as cave earth, but in other cases the result of burial can be a fossil that has turned literally to stone and is contained in a hard, cemented deposit of cave earth impregnated with calcium carbonate dissolved from the limestone.

Other suitable conditions for preservation can occur in areas where the bones have become incorporated in water-laid sediments, such as river channels, or lakeside sediments, as shown in figure 1.4. Indeed, one of the most impressive sets of fossils comes from the famous natural asphalt deposits of Rancho La Brea in California, where saber-toothed cats, lions, dogs, and their prey became trapped in the sticky material that lurked within pools of water, which proved to be an excellent preservative. The key to the process is therefore time and the right set of circumstances.

Caves can provide particularly fine specimens. Even when the deposit has become cemented, the bones can often be recovered by etching away the hard matrix with a weak (10–15%) solution of acetic acid, which dissolves the calcium carbonate of the deposit but leaves the calcium phosphate of the bone untouched. The specimen typically emerges in excellent condition and can be studied like fresh bone. With good material available from whatever source, measurements can be taken and the whole skeleton of the animal reconstructed where appropriate. From this we can observe details of size and proportion, estimate muscle masses and strength, and generally build up a picture of the living creature by integrating the fossil information with our knowledge of living relatives or broadly similar forms.

Teeth, which are no more than a very specialized part of the skeleton, also survive extremely well in suitable conditions. They have a complex structure, and consist of an inner bonelike substance known as dentine capped by a harder layer of enamel. The teeth, as we shall see, tell us much about the lifestyle of the animal, especially about its mode of feeding. Since they wear down in the process of chewing, they can also help us to establish how old the animal was at its death.

The Age of Fossils

All the animals that we are concerned with here have lived during the past thirty million or so years. But how do we know the length of time that has passed since they died?

It is possible to establish the relative ages of fossils simply by seeing if one deposit has been laid down after another. According to the geologists' "law of superposition," the latest deposit (with its fossils) is the youngest, and the assemblage of fossils permits us to correlate that deposit with others elsewhere. This principle has been understood for as long as people have been interested in the scientific study of fossils, and it remains an important element in paleontological interpretation. Biostratigraphy, as it is called, forms the basis of much of the geological time scale. But it does not, in itself, tell us how long ago the deposits actually formed. For this information we have had to await discoveries in other fields of science and the development of technology to permit their application.

The absolute age of the deposits containing fossil cats, and indeed the age of other rocks and of the earth itself, can now be established by a variety of methods. Many of the most useful techniques involve measuring the amounts of certain minerals that have been produced by radioactive decay from parent materials: by knowing the rate at which decay takes place, it is possible to estimate the time since formation of the deposit. These radioactive products are commonly found in deposits formed during volcanic eruptions, and in many parts of the world such deposits lie conveniently below and above rocks bearing fossils. Absolute dates are therefore usually

FIGURE 1.3 *Fossil formation in a cave*

The discovery of a partly articulated skeleton of the "scimitar-toothed" cat *Homotherium* in the late Pleistocene deposits of Friesenhahn Cave is the result of a series of fortunate circumstances. The spacious cave served as a den for the cats and other carnivores. In the first drawing we see our cat, an old individual as judged by the heavy wear on its teeth, seeking the cool shade beside the pool that occupied the center of the main chamber. The position in which the bones were found suggests that the animal simply lay on its side and died. The second drawing is a schematic view of the cave showing how alluvial deposits have entered and are about to bury the body. This must have happened shortly after the animal's death; otherwise, the remains would have been disturbed or even destroyed by scavengers. In the third drawing we see the cave several thousand years later. The alluvial deposits have finally choked the old cave entrance and buried the skeleton under layers of sediments that include the bones of other animals. In more recent times, one section of the cave ceiling has collapsed, and a sinkhole has opened to reveal the presence of a cavity and permit excavation of its contents.

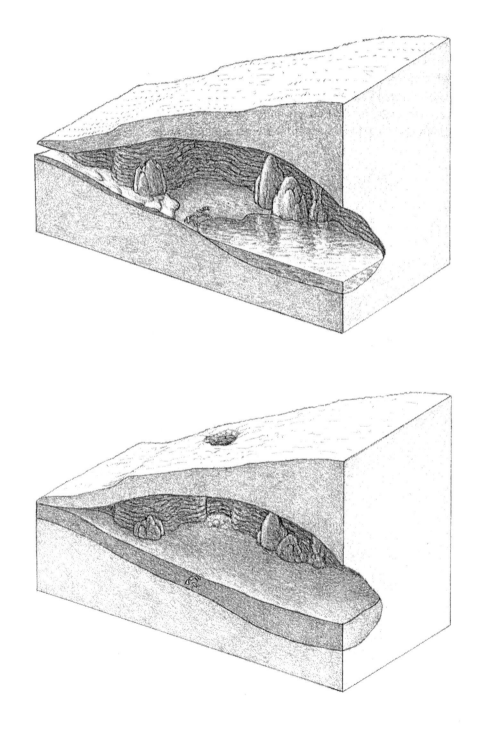

attached to the deposits in which the fossils occur, and rarely apply directly to the fossils themselves. The one well-known exception is radiocarbon dating, in which the ratio of radioactive to nonradioactive carbon is used to assess the age since death of the animal (or plant) itself. This ratio stems from the carbon in the air respired and the food ingested by an organism, and in all living organisms it remains constant; after death, however, it declines at a known rate. Unfortunately, the rate of decay is so rapid that the method can be used only for fossil animals and plants that died within the past forty thousand years or so—too short a time for most of the events that concern us here.

Of course not every deposit and its fossil assemblage can be dated accurately, but we still wish to know where to place the specimens in the order

of things. It is at this stage that we fall back on the principles of biostratig-raphy, which allow us to infer the broad age of a group of fossils in com-parison with similar animals or groups of animals from dated deposits else-where. Before the advent of absolute dating techniques, such a method was used to draw up chronological schemes, within which various deposits were placed in named eras and periods. Today we can place absolute ages on the boundaries between these divisions, but the names are still used extensively as a kind of working shorthand to refer to animals or events in broad time bands. Table 1.1 gives the names and ages for the most recent of the major divisions, those to which we shall make reference in the text.

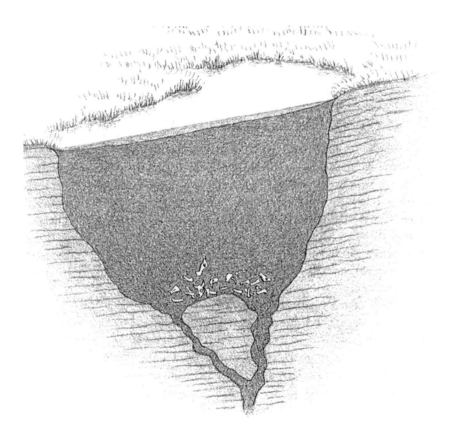

FIGURE 1.4 *Fossil formation in open-air deposits*
The illustration depicts the likely sequence of events that led to many animals being represented at the Rancho La Brea site in California. In the first scene a bison comes to the site to drink without realizing that the water lies on top of sticky asphalt, and the animal becomes trapped. In the second scene three indi-viduals of *Smilodon*, closely followed by three dire wolves, have been attracted by the stricken prey. In the final scene the bones of predators and prey are seen in a section through the deposits.

Table 1.1

Million Years (Ma)	Period
	Holocene (recent)
0.01	
	Pleistocene
1.6	
	Pliocene
5.0	
	Miocene
25.0	
	Oligocene
35.0	
	Eocene
55.0	
	Paleocene
65.0	

FINDING AND EXCAVATING FOSSILS

Fossils may be found in a variety of ways, from deliberate searches in likely localities and deposits to accidental discovery during the extraction of minerals or construction work. Wherever a river or a road-building project has cut through deposits of a suitable age, or where erosion by the sea has led to the loss of overlying sediment, fossils may well be found in the section or cliff that remains. Deposits in caves may contain hundreds of fossilized bones from a variety of animals. Since cats and other predators often seek refuge in caves, such deposits have proved to be a particularly rich source of information.

Recovery of the fossils, however they are discovered, requires careful excavation. Unless this is done, much information may be lost. Specimens will inevitably be damaged if handled carelessly, and bones from different individuals may become hopelessly mixed up. Fragile bones of young animals may be especially at risk, and if they are destroyed then our perception of the nature of the deposit can be totally altered. Above all, the context of the deposition may be obliterated, and with it a valuable guide to interpretation. The best course of action for anyone finding a fossil is to seek specialist advice before making any effort to remove items.

A proper excavation will recover the fossils slowly, permitting study of the position of the individual bones and ensuring that even the smallest items are found. Broken or damaged bones will be taken out in large blocks of sediment, or strengthened in place and then recovered, perhaps encased in a plaster jacket. Such methods inevitably require specialized equipment and materials and a good working knowledge of the anatomy of the animals with which one is dealing, in order to anticipate what may lie beneath or to the side of visible items. Fuller treatment may require laboratory facilities, repairing and strengthening broken or fragile items and ensuring that the specimens will not deteriorate with storage or handling. Permanent storage in a well-curated collection is the best way to deal with such remains, where controlled access will ensure that the fossils are available to future generations who wish to see and to study them in more detail.

RECONSTRUCTION

Finding and recovering the bones is only the beginning of the work that may be necessary to determine all that can be learned from fossil material. One of the most interesting aspects of study is the process of reconstruction, whereby the skeleton is first reassembled, with account taken of missing items, and then an effort is made to depict the once-living animal in its surroundings.

Here the most important aim is to integrate the information from the fossil and its associated deposits with that obtained from closely related or analogous living organisms in similar situations. A knowledge of functional anatomy helps to put the flesh, and eventually the skin, back onto the bones, while the information about the environment of the time and the range of other animals permits us to show the living conditions.

The illustrations in this book have been based on the skeletal shape and proportions of the species in question, or on reasoned approximations in the case of damaged or missing pieces. We discuss the process in detail in chapter 4, where the anatomy and function of the cats is considered, but an overall impression of the process is given in figure 1.5.

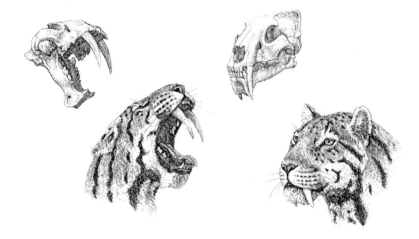

FIGURE 1.5 *A brief illustration of the principles of reconstruction based on the head of the dirk-toothed cat* Megantereon

As may be seen in these sketches, the known bony anatomy—or osteology—of fossil cats is the most important basis for reconstructing their life appearance. Soft tissues are then added using living species as a reference and any markings on the bone as a guide to muscle mass and strength. For more detailed discussion see chapter 4.

CHAPTER 2
Evolution and the Origin of the Felidae

AMONG THE VAST ARRAY OF FORMER LIFE FORMS, EVERYTHING from jellyfish to dinosaurs, the Felidae have a reasonably good fossil history. Other things being equal, the chance of survival of an animal through the sequence of death and burial to discovery as a fossil is largely dependent on its size, and many of the cats have been fairly large individuals. Of course, as large predators, the cats occupy a very particular place in the web of life, one that may be most succinctly described as the apex of a pyramid of numbers. In order to survive, any organism must be outnumbered or outweighed by its food source, which is why antelopes and zebras will always outnumber lions or spotted hyenas while requiring extensive areas of grass for fodder. It is estimated that the Carnivora, which make up approximately 10% of all mammalian genera, account for only about 2% of the biomass of mammals.

The fossilized remains of cats are therefore generally scarcer than those of many other species. Fortunately, the natural imbalance in living numbers has to some extent been overcome in the fossil record by the prevalence of cave deposits and natural traps. As we pointed out in the last chapter, the evident use of such caves as lairs and eating sites by many predators has meant that the remains of the hunters as well as the hunted are often discovered in these sites, which have functioned as natural concentrators of bones. However, it is usually bears and hyenas that are the most abundantly represented of the larger carnivores, whereas the cats are more often represented by only a few individuals. Some of the major exceptions to this occur in caves that seem to have acted as traps into which the cats may have fallen or jumped when tempted by the carcasses of other victims. Equally important as find localities (and in some cases even more so) are a number of open-air sites, such as the Rancho La Brea asphalt deposits, where cats and other predators became fossilized vastly in excess of their natural densities. It is by piecing together the evidence from a series of localities that we gain much of our information about the evolution of the family.

If we glance over the fossil history of the Felidae we see a range of animals that appear broadly catlike in their teeth and skeletons, although on closer inspection many of them turn out to be rather different from the species that we know today. Of course, only the bones and teeth generally tend to remain in the fossil record, so that strictly speaking we are usually talking about differences in those features rather than in the complete animals—but when examined in the light of what we know about living species, this evidence often permits a good deal to be said. Moreover, as we have already mentioned, fossilized footprints may survive, and in exceptional circumstances so too may evidence of soft tissues. The German sites of Messel (Eocene) and Höwenneg (Miocene) have produced skeletons with traces of hair and stomach contents, for instance, although to our knowledge no cat remains have been found in such condition.

The link between the past and present is evolution, the organizing concept of biology. It is evolution that has produced the vast array of plants and animals that we see around us, and we begin with a brief summary of some of the major points in order to set the appearance of the Felidae in its proper perspective.

EVOLUTION

Life as we know it has evolved over millions of years as a result of natural selection operating on the variation that occurs between individuals. This has been clearly seen since at least the time of Darwin's publication of *On the Origin of Species by Means of Natural Selection* in 1859. But why is that variation there, by what mechanism is it produced, and how do new species arise?

Evolution as a field of study is an enormous and daunting topic, and yet the key points, like the mechanism for inheritance, are essentially quite simple. The genetic code contained within the molecules of DNA (deoxyribonucleic acid) in the chromosomes of our cells, and passed on from parent to offspring, lays down the basic plan that the individual will follow during growth and development. The code is a simple one, yet it is capable of transmitting a vast amount of information. In sexually reproducing species like ourselves and other mammals, the offspring receive genetic information from both parents, information that each parent has in turn received from its own parents. The genetic information passed on to the next generation by each parent differs with *every* sperm or egg produced, since the genetic information originally received by that parent from the previous generation is differently combined in each egg or sperm. The effect is that, with the exception of identical twins formed from a single fertilized egg, each offspring differs genetically from its parents and even from its siblings (although family characteristics may still be seen). The result is a virtually infinite range of possible variation within an overall plan of organization for each species—a range that permits us to recognize individuals of our own species by the differences in size, shape, and facial structure, for example, even though we all have the same generalized features. It is on the variation in characters of the organism—such as strength, endurance, speed, and so on—that natural selection has operated over vast spans of time to induce directional change, and to fit organisms to their environment. Note, however, that the fitness does not have to be perfect: survival to reproduce is the key test of fitness.

In plants and animals that reproduce sexually, the genetic information from each parent that is necessary to produce the next generation is combined in the manner described above only by organisms that recognize each other as mates. Cats are no different in this respect. The species, be it a lion, a tiger, a leopard, or a parrot, is simply a group of organisms bound together by a shared system of mate recognition and fertilization. That system will involve signaling and response between potential mates, but the precise form of the system will vary between species, between families, and between orders. Some use calls; others, coat color and patterning; and others again may use movements and display—and all may be employed to some extent in complex courtship rituals. In some species, chemical signals may be of equal or even greater importance, but the precise nature of the fertilization system will depend on the habitat within which the organism exists.

We think of natural selection as a force for change, fine-tuning an organism to the environment in which it lives. Over time, teeth become better adapted to coping with food, and legs become better fitted for running, for climbing, or for digging. The cats appear to bear out this interpretation: they are superbly equipped to hunt, catch, and kill their prey, and their teeth make short work of the carcass. In general this view is undoubtedly correct, but natural selection itself is also an intensely conservative force: it will operate against extremes in characters, such as teeth that are too large to permit proper feeding, and it will act on the components of the fertilization system to ensure that they remain stable while the organism remains in its normal habitat. While some variation in tooth structure will not matter too much in processing food, and legs that are a little shorter may still carry an animal far enough or fast enough, the system has to operate with some precision or there will simply be no next generation however well adapted the animal may be in some ways.

The species will therefore persist while conditions are constant and the fertilization system remains stable. But should conditions change, then the population may find that the new circumstances affect its lifestyle. In that case, the population has three choices: it may adapt to those conditions, it may become extinct, or it may move to a more suitable area. If the changed conditions leave a small part of the original population isolated, and if those new conditions affect the fertilization system, then natural selection may operate relatively rapidly to change the system in that small population. A simple example would be a change to a more wooded environment, where visual signals between mates might be less effective, and where selection for a system of calls might operate. If the group with the new system should now meet up with the original population again, mating *may* be impossible as a result of the differing fertilization systems. At that point a new species has originated, since the two populations do not share the same fertilization system. Such new species, because they have only a portion of the original pool of inherited variation, may tend to differ relatively rapidly in morphology from the ancestral species, and may be recognizable to us in the fossil record.

In this way we can understand two of the major characteristics of the evolution of life seen in the fossil record: the development of features that fit organisms to their environment, and the appearance of new species. We should also note two other points. The first is that the ancestral species may very well continue to exist after the event that gives rise to the new species. The second is that speciation, extinction, and dispersion are all related effects. It is probable that most of the major alterations in the development of life—those involving dispersions, extinctions, and the origin of new species—have been directly provoked by changes in the physical environment. In the absence of such changes the species tend to persist, which is why many species, particularly those with wider tolerances of changed conditions, have tended to persist for so long.

We can understand the mechanisms of evolution by examining living organisms, armed with increasing technological refinements that permit us to study the building blocks of life in the laboratory in ever more detail. It is this approach that has permitted us work out the genetic code and to see how it is copied from cell to cell and from generation to generation. Such studies are difficult, and the laboratory techniques to be mastered present a formidable barrier to progress and to easy understanding by the nonspecialist. But the long-term evidence of evolution at work over vast spans of time can be seen by all in the changing patterns of the fossil record. The fossil history of the cats should be seen as simply one part of the complex picture of the evolution of life on the earth.

Origins of the Carnivores and the Appearance of the Palaeofelids

Individual fossils may survive for millions of years in excellent condition, but, as one might expect, the further back in time we go the less detailed is the information we obtain from the fossil record. The sediments that held the fossils may have been destroyed, or there may simply be gaps where suitable deposits never formed. We know that anything alive today must have had ancestors, and we can recognize animals that were clearly members of the Carnivora from around 60 Ma ago (in other words, during the Paleocene). But cats as we understand the term (Neofelids, or modern cats, as they are sometimes called) are simply not known with certainty until around 30 Ma ago, in the middle of the Oligocene, and their fossil record only really improves around 10 Ma ago, toward the end of the Miocene. In the intervening period we see a number of animals that are clearly catlike in their morphology, which were formerly considered by many paleontologists to be ancestral to the true cats; as such, they were referred to as Palaeofelids, or ancient cats.

However, more recent studies now suggest that these catlike animals differ enough from the modern cats to be considered a separate family, the Nimravidae, as summarized by Harold Bryant. Chief among the skeletal characters used in making this distinction is the formation of what is known as the auditory bulla (figure 2.1), a capsule that houses the middle ear mechanism of malleus, incus, and stapes, the bones that connect the eardrum to the inner ear and thereby transmit sound to the brain. In ourselves and other members of the Primates, this mechanism is housed within the base of the skull; but in many other orders, including the Carnivora, an external bulla is the rule. In the true cats the interior of the bulla is separated into two chambers by what is termed a septum, but nimravids lack either the septum or the entire bulla, implying only a cartilaginous housing for the ear mechanism that has not fossilized. The Felidae are therefore distinguished by what is termed a shared character with respect to the condition of either no septum or no bulla. In this interpretation, the Nimravidae would have evolved alongside the earliest ancestors of the true cats, and would not themselves be ancestral to the Neofelids.

The diagram in figure 2.2 shows the Nimravidae in such a position, as one more family in the phylogeny (the pattern of ancestor-descendent relationships) of the Carnivora. If that separation is correct, then the Nimravidae exhibit a condition known as convergence in the degree to which they have developed, in parallel, some of the specialized morphological features of the true cats. These features include both the general skeletal features that we associate with the cats, such as a long-limbed appearance ending in feet with claws, and characters of the skull and dentition that include shortening of the face, the development of slicing teeth, and the appearance of large and pointed canines. It is in these latter characters that the parallels with the true cats are most marked, especially in the development within some genera of quite enormous upper canines into what are commonly known as "sabers." These teeth are long, flattened, and curved, and the effect on the appearance of the skull can only be considered bizarre to a modern observer.

Various nimravid species have been identified and placed in a number of genera, shown here in a phylogeny of the Nimravidae in figure 2.3. Note

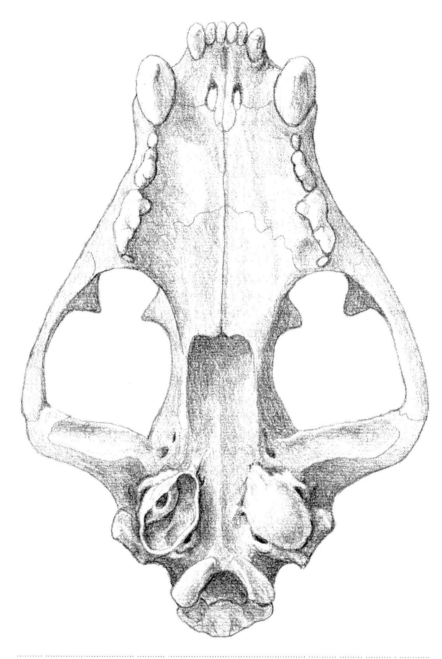

FIGURE 2.1 *The structure of the auditory bulla*

In this ventral view of the skull of a tiger, we have shown the left bulla intact and the right bulla sectioned to reveal its structure. Inside the chamber is a second, smaller one. The wall separating the chambers is called the *septum*, and it has also been sectioned to show its hollow interior.

In the catlike nimravids these structures are different: in some nimravids the bulla was cartilaginous, and is therefore not preserved in the fossils. In others, the bulla is ossified but lacks a true septum, although there may be a superficially similar structure called the *proseptum*, which is formed of bones different from those that make up the septum of the true cats.

how these animals are placed in this scheme in tribes, reflecting opinions about the closer relationships of some genera within the family. One of the nimravids, *Hoplophoneus occidentalis*, was about the size of a large leopard and had only moderately increased canines (plate 1). In contrast, the somewhat larger *Hoplophoneus sicarius* and *H. mentalis* (figure 2.4) and the lion-sized *Barbourofelis fricki* (figure 2.5) had huge upper canines, matched by a massive flange at the front of the mandible. In the latter species, the deciduous or milk canine that preceded the permanent upper canine was also so large that it did not erupt into position until the animal was already quite well developed and the permanent cheek teeth (the dentition behind the canines) in wear, since there simply was no room in the face for the tooth. This delayed sequence of tooth appearance must have had a significant effect upon the abilities of the animal to deal with prey, and has led to suggestions that the period of parental care was extended as a result, as suggested in figure 2.6.

In contrast with these large-toothed forms, another nimravid lineage led to the appearance of a more "normally" catlike species known as *Dinaelurus crassus* during the later Oligocene. This species shows particular parallels with the specialized morphology of the cheetah, in view of its markedly shortened face, highly domed skull, and small canine teeth.

We should not be misled into thinking that the nimravids were somehow inferior to the true cats. As a family, their appearance predates that of the Felidae and lasts for almost as long in the fossil record. Nimravid diversity and dispersion suggest that by any objective criterion they were a successful group of animals during the Oligocene (figure 2.7) and the Miocene (figure 2.8).

It is worth stressing that the parallel evolution of the Felidae and Nimravidae is simply one example of convergence among meat-eating animals. Two other major instances may be highlighted. The first is the order Creodonta—unrelated to the Carnivora, but the dominant group of carnivorous animals over much of the world between the Paleocene and the Miocene. Between the two families Oxyaenidae and Hyaenodontidae, the former rather catlike and the latter rather dog- and hyena-like, the creodonts spanned the size and morphological ranges of the true Carnivora. Two of the catlike genera in particular, *Apataelurus* and *Machaeroides*, presaged the development of saber-toothed forms among the Nimravidae and Felidae in the detailed development of their dentition and skull characters, as shown in figure 2.9.

The second group is the carnivorous marsupials of Australia and South America, continents long isolated from the rest of the world during the past 65 Ma, where unique faunas were able to evolve. Although even more distantly related to the Carnivora, the marsupial carnivores again span the range of morphology seen in the families of the placental order. Perhaps the most remarkable from the point of view of the subject of this book are

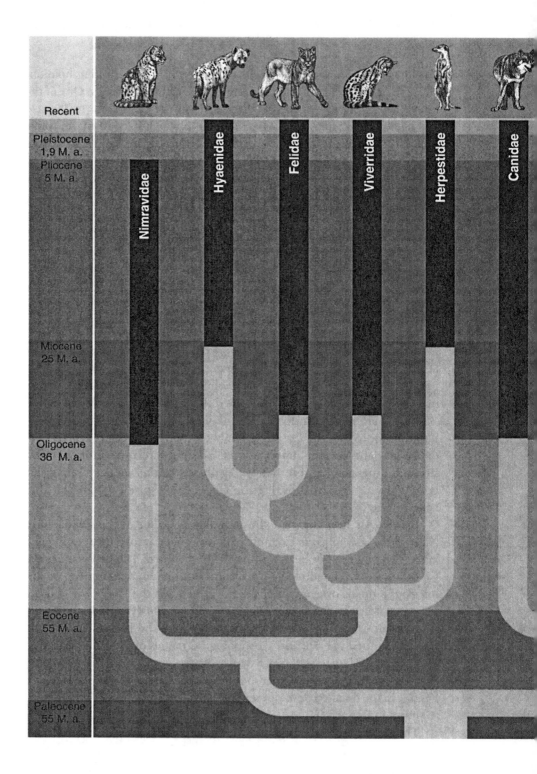

Recent

Pleistocene
1,9 M. a.

Pliocene
5 M. a.

Miocene
25 M. a.

Oligocene
36 M. a.

Eocene
55 M. a.

Paleocene
55 M. a.

Nimravidae

Hyaenidae

Felidae

Viverridae

Herpestidae

Canidae

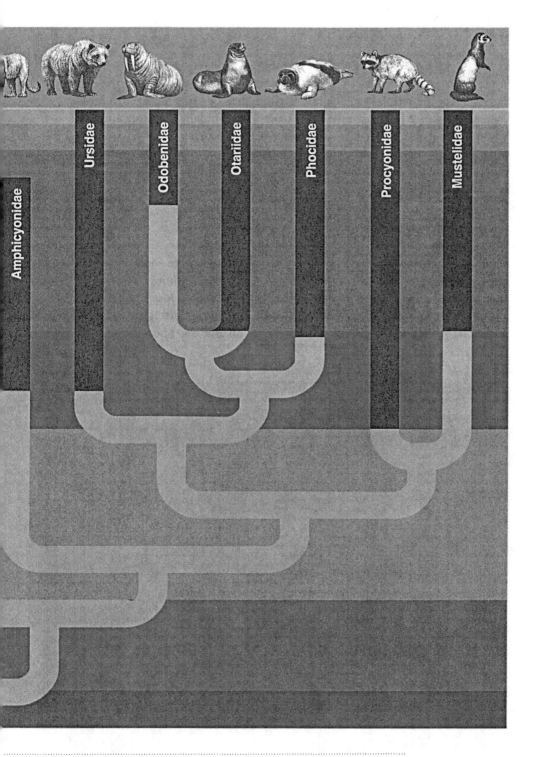

ƒURE 2.2 *A phylogeny of the Carnivora*

ny such scheme should be seen as a hypothesis of relationships based on current
nderstanding. Differences of opinion about the pattern of ancestry and descent
ll exist between workers.

the spectacular developments in saber-toothed, catlike animals such as *Thylacosmilus* in South America, shown in figure 2.10. In Australia, the largest of the marsupial carnivores was the strange "marsupial lion," *Thylacoleo* (figure 2.11), an animal with closely set caninelike incisors and extremely elongated premolars with an impressive shearing action. Marks on contemporaneous bones precisely matching the unique shape of the cutting edges of these teeth have established beyond reasonable doubt that these animals ate meat.

THE APPEARANCE OF THE NEOFELIDS

With the recognition that the Nimravidae constitute a separate family from the Felidae, the origin of the latter is less clear than was once thought. What remains reasonably apparent is the fact that felids are more closely related to hyenas, herpestids (mongooses), and viverrids (civets), and these four families are placed together in the suborder Feliformia, also known as the

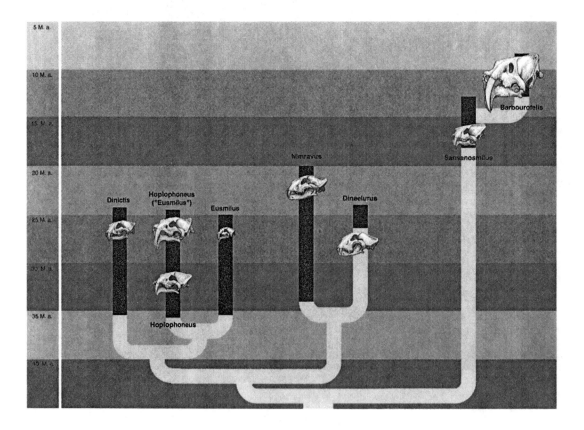

FIGURE 2.3 *A phylogeny of the Nimravidae*

aeluroid carnivores; while canids, together with mustelids (which includes otters, badgers, and weasels), procyonids (raccoons), and ursids (bears), are grouped under Caniformia. Within the latter, incidentally, are placed the seals, walruses, and sea lions, and although these do not concern us here their relationship to the terrestrial carnivores should be borne in mind. These relationships are illustrated in figure 2.2.

The first of the modern or true cats are placed in two genera. The earliest, found in French deposits with dates centering on 30 Ma ago, is known as *Proailurus lemanensis*, a smallish animal with a skull perhaps 15 cm in length (figure 2.12). Although the shape of the skull and the form of the teeth are both broadly similar to those of modern cats, the skull contains more teeth—a primitive feature in comparison with the reduced number of teeth seen in the dentition of more recent fossil specimens and in living species (figure 2.13). In the New World, the earliest known cat is currently the proailurine reported by Robert Hunt from Nebraskan Miocene deposits of around 16 Ma ago.

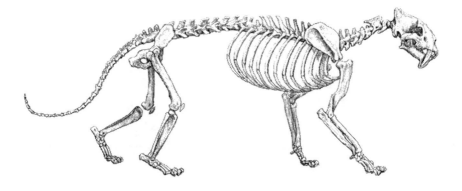

FIGURE 2.4 *Skeleton of* Hoplophoneus mentalis

This very early species of the genus comes from the Chadron (late Eocene) Formation of the western United States. Despite its great age this nimravid "cat" is in many respects very specialized, especially in the development of its upper canines and mandibular flange and in the overall robustness of its build. These derived characteristics resemble those of the felid genus *Smilodon*, but in *Hoplophoneus* we also find a much longer back and metapodia that are relatively shorter than in even the latest species of *Smilodon*. The long back seems to be a generally primitive character among carnivores, while the reduction in the metapodia of *Smilodon* implies a secondary trend away from longer-footed ancestors.

The humerus, or upper forelimb bone, shows very prominent areas for insertion of the deltoid and supinatory muscles, and in this feature *Hoplophoneus* resembles later barbourofelines and thylacosmilids. Smilodontines derive from "normal cats" and are thus less extreme in these traits.

Reconstructed height at shoulder 48 cm.

We consider such an arrangement of teeth to be "primitive," and use that term, because it resembles the ancestral condition seen in earliest mammals from which later forms are descended. These ancestral animals had a much larger number of teeth than most living species. As the various lineages have evolved, some teeth have tended to become specialized and more important, while others have become reduced and even lost. We refer to such a dentition as more "derived," or "advanced," but there is no implication of inferiority or superiority in any of these terms.

From about 20 Ma ago we have evidence of cats assigned to the genus *Pseudaelurus* (figure 2.14). The members of this genus are thought to have been ancestral both to the modern, living cats and to the well-known saber-toothed species, also known as machairodonts. We may therefore envisage a split in the tree of relationships at that point, with one line, placed in the subgenus *Schizailurus,* leading through to the living and fossil conical-toothed species of true cats; and another, placed in the subgenus *Pseudaelurus,* lead-

FIGURE 2.5 Barbourofelis fricki *with a flehmen gesture*
As perhaps the most extremely developed of all saber-toothed carnivores, *Barbourofelis fricki* must have been a very impressive sight when baring its teeth in the flehmen gesture. Apart from their use as hunting weapons, the enormous canines of these nimravids were probably important for display during encounters with other members of the species.

ing to the now extinct saber-toothed species, the machairodonts. The latter lineage is typified by one species in particular, *Pseudaelurus (Pseudaelurus) quadridentatis* from middle Miocene sites at Sansan in France and Buñol in Spain, which exhibits machairodontine features. In contrast, other taxa, such as *Pseudaelurus (Schizailurus) transitorius* and *P. (S.) lorteti*, are more gracile (slender) and look much more like ancestors of modern cats (figures 2.14 and 2.15).

The relationships of the Felidae are shown schematically in figure 2.16. The living and fossil conical-toothed species after *Pseudaelurus* are given subfamily status as Felinae, while the extinct saber-toothed subfamily Machairodontinae is divided into three tribes. The first of these, the Smilodontini, includes the well-known American saber-tooths of the genus *Smilodon*, as well as the widely distributed genus *Megantereon*. We have also accepted here that it contains the Miocene genus *Paramachairodus*, a cat the size of a small female leopard that some have suggested may have been the ancestral species of the other later and larger species. The tribe Homotheriini contains genera such as *Machairodus* and *Homotherium*, while the third tribe, the Metailurini, contains genera such as *Dinofelis* and

FIGURE 2.6 *Family group of* Barbourofelis fricki *attacking a rhinoceros*
The amphibious, hippolike rhinoceroses of the genus *Teleoceras* were abundant herd animals of the American Miocene, and a tempting source of food for any predator strong enough to attack them. The adult female at left tries to force the rhino to roll onto its side, and encourages the large cubs to help. The cubs, more than a year old, are of nearly adult size and weight, but although their size is of some help in such a contest only the mother will be able to kill the rhino, probably by slashing at the abdomen to induce massive bleeding (see chapter 4). The cheek teeth of the cubs are fully functional, but the milk canines are only just erupting and the flange on the lower jaw is just beginning to develop. Such cubs will therefore be dependent on their mother until well into their second year.

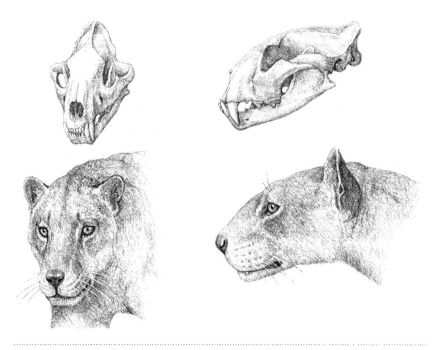

FIGURE 2.7 *Studies of the head of* Nimravus brachiops

This species is known on the basis of a fine series of skulls from Oligocene deposits in North America, mostly from Oregon, Nebraska, and South Dakota. Little is known about its postcranial skeleton, but it seems that the limb bones, and the bones of the feet in particular, were more slender than in other nimravids of the Oligocene such as *Hoplophoneus*. The proportions of the skull are strikingly catlike, and in the flesh the animal would have looked more familiar to us than many of the fossil species of true cats. However, a closer inspection of its teeth and of its ear region show that it is a nimravid—indeed, the genus has given the name to the family.

FIGURE 2.8 *Sequential reconstruction of* Barbourofelis morrisi

This is one of the three species of the genus *Barbourofelis* recognized in Clarendonian and Hemphilian deposits of Miocene age in North America. A reasonable number of complete long bones have been recovered, but no complete skeleton is known so that relative proportions can only be estimated. For these estimates the large sample of remains of the similar species *Barbourofelis lovei* from Florida, described by Jon Baskin, provide an important frame of reference.

These nimravids were very powerfully built, with short distal limbs and strong muscle insertions, and the claws were hooded and retractable as they are in modern felids. The vertebral column is poorly known, and although our reconstruction shows it stronger than in primitive nimravids more data are necessary to confirm this interpretation.

The muscle masses of the animal (drawing *2*) would be somewhat intermediate between those of a large cat and a bear. Like bears it had enormously developed extensors and flexors of the forepaws, as well as very powerful deltoids. Our depiction of the living animal (drawing *3*) with a mostly plain coat reflects the idea that it may have dwelt partly in open habitats, while the markings of the face and neck are there for communication with other members of the species.

Reconstructed height at shoulder 65 cm.

while the third tribe, the Metailurini, contains genera such as *Dinofelis* and *Metailurus.*

The diagram shows that most of the felid taxa that we know are therefore confined to the past ten or so million years. We discuss the membership of these categories, and provide further information about many of the species, in the next chapter. For the present, the diagram of phylogeny should give some overall idea of the diversity and broad pattern of relationships within the Felidae. It is clear, for example, that the saber-toothed cats are not ancestors of the living cats, any more than gorillas or chimpanzees are ancestral to us humans. Both subfamilies of cats evolved in parallel from earlier ancestors, in much the same way as the living great apes and humans.

We may as well clear up one point at once and say that while the term "saber-tooth" (from *saber*, a curved sword often used by mounted cavalry) describes some features of the dentition of these cats well enough, there is simply no basis for calling them tigers: they are not closely related to the

FIGURE 2.9 *Life reconstruction of the creodont* Machaeroides eothen
Machaeroides eothen is well known from associated cranial and postcranial material from the Bridger (middle Eocene) Formation of the United States. The animal was rather small, around 30 cm at the shoulder, and the limb bones are strikingly robust for a creature of its size. The elongation of the canines and the associated cranial features expected in the machairodonts are not well developed, but other species such as *M. simpsoni* exhibit them quite clearly. The overall proportions suggest that members of the genus *Machaeroides* were powerful little animals, and the bones of the similarly sized nimravid *Eusmilus cerebralis* appear almost fragile in comparison.

true tiger and are certainly not an ancestor of that species, and there is no evident reason to assume that they had striped coats. Nor should we make the common mistake of assuming that because they no longer survive, such saber-toothed cats were inadequate animals, in some way necessarily inferior to the living species in general. In fact these animals were highly specialized in their dental features, and had merely taken a different evolutionary pathway. Many of them were successful species that existed over wide areas of the world for several millions of years, and in eventually becoming extinct they simply shared the fate of most forms of life that have ever existed on the earth.

Despite the outline understanding of felid evolution that we now have, it should be stressed that we still lack any clear idea of the immediate ancestry of the living species or of the more precise pattern of relationships between them. The typical conical-toothed cat of the later Miocene of Europe, for instance, is a small species known as *Felis attica*, presumably the stock from which the living species evolved. Of course the fossil record can

FIGURE 2.10 *Life reconstruction of* Thylacosmilus atrox

This marsupial saber-tooth was described in 1934 by Elmer Riggs on the basis of two partial skeletons found in Pliocene deposits in Argentina, and this material remains the most complete evidence for the species. The animal was about the size of a South American jaguar and had the same short, heavily muscled limbs that suggest hunting by ambush; unlike the placental saber-tooths, however, it lacked retractable claws, and this has led to suggestions that it must have hunted in a very different manner. But the articulations and muscle attachments of the forelimb clearly point to a developed grasping ability, and it is likely that *Thylacosmilus* used its front paws to subdue and hold prey in a broadly catlike manner. The living cuoll (*Dasyrus quoll*) of Australia, often known as the marsupial cat, is a good example of a hunter that does comparatively well using catlike techniques in spite of not having retractable claws. These examples show us how parallel evolution in unrelated animals can produce similar outcomes. See chapter 4 for a discussion of claw mechanisms and for a comparison of the head and neck of *Thylacosmilus* with the saber-toothed cat *Smilodon* (figure 4.30).

Reconstructed height at shoulder 60 cm.

FIGURE 2.11 *Life reconstruction of* Thylacoleo

The anatomy of this strange animal has puzzled paleontologists for decades, and some have seriously questioned whether it was actually carnivorous. Recent studies of its masticatory apparatus, together with the discovery of bones bearing marks that precisely match the shape of its peculiar cheek teeth, show beyond reasonable doubt that it was, but its overall structure betrays its origins among the phalangerids, a group of rather specialized arboreal and largely vegetarian marsupials that include the living phalangers and Australian opossums. *Thylacoleo* developed a much greater size, longer limbs and neck, and a shorter back—proportions much more in keeping with a terrestrial way of life, which have led to comparisons with the lion. However, the appearance of the living animal must have been unique.

Reconstructed height at shoulder 65 cm.

be expected to tell us only so much even in perfect conditions, and the precise lineage linking any one species to another, particularly as ancestor and descendant, is a matter of inference if not speculation. An alternative approach much favored by many interested in systematics is to examine the biomolecular structure of the living species as a guide to patterns of relationships. The basic premise of such methods is that differences and similarities in molecular structure should point to similar relationships between extant species. Moreover, it is argued, the amount of difference between closely related species should be proportional to the length of time since they diverged from a common ancestor, and should therefore permit some estimate of the time of that divergence.

The problem with these methods is that the interpretation of results is never easy. Most of the studies undertaken provide a reasonably sensible pattern of relationships congruent with what we already believe from more conventional morphological analyses (although one recent effort based on an analysis of material preserved in fossil bones suggested a close link

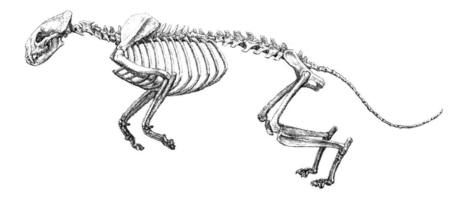

FIGURE 2.12 *Skeleton of* Proailurus lemanensis

The skull and most of the limb skeleton of this very early felid are known from finely preserved limb bones belonging to two individuals found at the Oligocene locality of Saint-Gérard-le-Puy in France. Other sites have yielded important although less complete specimens, but the vertebral column remains largely unknown. Here we have restored *Proailurus* on the basis of its likely Miocene descendant, *Pseudaelurus.*

The bones of *Proailurus* are very similar to, if slightly larger than, those of the living viverrid *Cryptoprocta*, the fossa of Madagascar, and it seems likely that the early cat was just as adept at climbing and jumping from branch to branch.

Reconstructed height at shoulder 38 cm.

FIGURE 2.13 *Mandibles of* Proailurus *(top),* Pseudaelurus *(center), and a modern cat to illustrate the teeth*

Notice that the teeth lost in modern felids were already quite reduced in *Proailurus,* and that further loss of teeth was achieved by the saber-toothed cats.

between the American *Smilodon* from the asphalt deposits of Rancho La Brea and modern cats such as the leopard and lion that is clearly invalidated by the known fossil record). However, many of the suggested times of separation published to date do not sit particularly well with what we know from the fossil record. We may not know of the earliest appearance of a given species, for instance, but we can at least often say whether it is found as a fossil before it is supposed to have appeared. Moreover, placing a time on separations between species in even the most sensible biomolecular classifications requires that the biological "clock" of molecular divergence be somehow calibrated against events of known date and then assumed to operate in a linear fashion. The first requirement can be difficult to achieve; the second is open to question, and it would seem that we still have some way to go in understanding the chronology. This point is considered further in chapter 3.

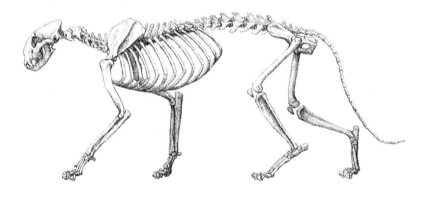

FIGURE 2.14 *Skeleton of* Pseudaelurus lorteti
This cat, about the size of a large lynx, has much in common with modern cats. In fact, individual limb bones of *Pseudaelurus* (*Schizailurus*) can hardly be distinguished from those of modern species. But the proportions of the pseudaelurine cats were still partly reminiscent of some viverrid-like ancestor: the back was longer than in living animals, and the distal segments of the limbs did not exhibit the elongation characteristic of modern felids.

Reconstructed height at shoulder 48 cm.

FIGURE 2.15 Pseudaelurus *sp. jumping up a tree.*
This reconstruction is based on a partial skeleton from New Year Quarry, a Barstovian (early Miocene) site in California. In size and proportion this animal was very similar to a small puma, the main difference being in the shorter metapodia of the fossil species. Such proportions point to a very able climber. A very similar, if slightly more gracile, skeleton was recovered from younger deposits of Clarendonian age in New Mexico.

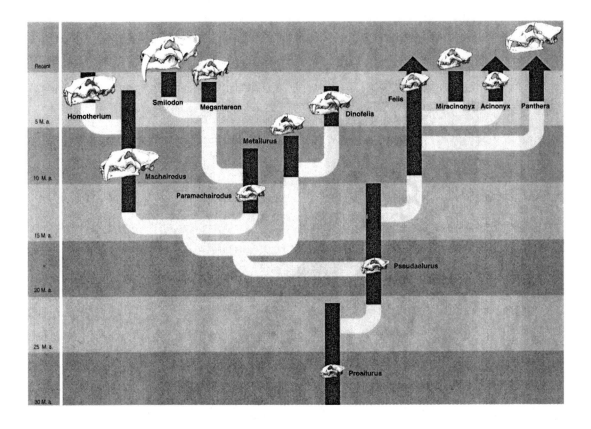

FIGURE 2.16 *A phylogeny of the Felidae*
See text for further explanation.

CHAPTER 3
Individual Species

I N THIS CHAPTER WE INTRODUCE AND OUTLINE WHAT WE KNOW OF some of the most interesting and significant of the larger cats, living and extinct, giving some details of their size and their known or inferred distributions. We have already outlined the principles of systematic arrangement in chapter 1, where we discussed the various categories within which animals are usually organized for scientific study. We follow such an arrangement here, because it has the very distinct advantage of providing a grouping for the species that is logical and based on relationships. Otherwise we might as well list the species alphabetically; that would have its own logic, of course, and for certain purposes it would be adequate—but not if we want to discuss species on the basis of their shared ancestry.

One point should be stressed at the outset. As we mentioned in chapter 1, the history of the nomenclature of the cats, and the interpretation of relationships, has at times been extremely confused, with different authors holding widely divergent opinions on generic allocations and relationships. This is true even of the living species, and is especially so in the case of many of the fossil taxa even today. Because of the past confusion we make no effort to deal here with every fossil species of larger cat named in the literature, and this is certainly no place to attempt a revision of the nomenclature (any such undertaking, even if possible, would produce far more information than necessary for the purpose of this volume, and would tend to confuse, rather than to enlighten, the general reader). Instead, we have been selective in an effort to present a coherent picture, making only passing reference at times to other taxon names that may be encountered by anyone interested in reading more widely. Where more than one species of a genus is known, as is usually the case, we group them for convenience under the heading of that genus. What we present here is therefore our own synthesis of the information; anyone interested in pursuing the topic further should consult our list of suggested further reading.

Since we have already discussed the earliest-known true cats of the genera *Pseudaelurus* and *Proailurus* we shall not repeat that information in this chapter, although we consider them again in chapter 4 when we look at the anatomy and function of the cats. Here we shall concentrate on the two subfamilies—the saber-toothed Machairodontinae and the conical-toothed true cats, the Felinae—that appear in the fossil record after *Pseudaelurus* (and, in particular, during the past ten million years or so).

SUBFAMILY MACHAIRODONTINAE

Tribe Metailurini

The composition of this tribe continues to be a particular source of confusion, in large part because of the fragmentary nature of the evidence. Specimens that look broadly similar have at times been variously allocated to the genera *Paramachairodus*, *Pontosmilus*, *Adelphailurus*, *Stenailurus*, *Metailurus*, and *Therailurus*, although *Therailurus* is now thought to be a junior synonym (a term meaning a later name for the same taxon, where the earlier name generally has priority) of *Dinofelis* (see below). The genus *Pontosmilus* alone has had four species (*P. ogygius*, *P. hungaricus*, *P. schlosseri*, and *P. indicus*) referred to it by some authorities, including at least one species that was previously referred to the genus *Paramachairodus*. However, it is by no means clear that the genus *Paramachairodus* does in fact belong within this tribe, and we have therefore elected to follow opinion that places it within the tribe Smilodontini, perhaps as an ancestor of the dirk-toothed cats *Megantereon* and *Smilodon* (see below).

The specimens referred to the tribe Metailurini are of later Miocene–earliest Pleistocene age, with a mostly Eurasian distribution, and the majority are about the size of a modern leopard, with the upper canines moderately elongated and flattened. However, no complete skeletons or even complete skulls are as yet known and published for many of these proposed taxa, which makes reconstruction rather difficult. Some authors have seen fit to place the entire tribe in a separate subfamily, the Metailurinae. In large part this disagreement stems from the fragmentary nature of the material and a great deal of overall similarity (although some forms appear more primitive in their retention of extra upper teeth, as in the case of *Adelphailurus* and *Stenailurus*, which retain an upper second premolar). More work and more complete materials are clearly needed to disentangle the complex pattern of relationships.

ADELPHAILURUS. The type material of *Adelphailurus*, a fragmentary skull and dentition assigned to the species *A. kansensis*, comes from Hemphilian-age deposits of the Edson Quarry in Kansas (figure 3.1). These fossils, together with unassociated postcranial remains referred to the same

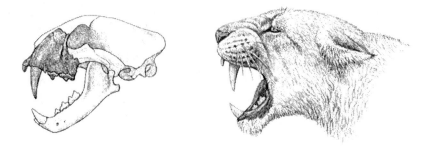

FIGURE 3.1 Adelphailurus kansensis *from Kansas*
Reconstruction of the skull and head based on the type specimen UKM 3462 from the Edson Quarry, Sherman County, Kansas.

species, suggest an animal of puma size. The material is of particular interest for the light it may shed on a rather enigmatic fossil from the South African early Pliocene site of Langebaanweg in Cape Province. This specimen, a maxillary fragment from an animal also about the size of a puma or a small leopard, was assigned to *Felis obscura* by the discoverer, Brett Hendey, conveying his lingering uncertainties about its true identity. More recently, the American paleontologist Dan Adams has suggested to us that it might be rather similar to *Adelphailurus*—and indeed the similarities to the Kansas type are marked, despite the fact that the American specimen retains a small second molar missing in the South African cat.

METAILURUS. This genus, together with the various species of the genus *Dinofelis*, is one of the best-represented members of the tribe. *Metailurus major* (plate 2) was based originally on material found in China. Although the Chinese specimens are not particularly well dated, European fossils of this species from Samos, Halmyropotamos and the Teruel Basin are all of Turolian land-mammal age, about 8.0 Ma. The animal had what are termed advanced features in the loss of the second upper premolar (a term used, as we have stressed, because we know that the ancestral cats had higher numbers of teeth and that such reductions occur during the evolution of lineages). It also had moderately elongated upper canines.

DINOFELIS. The genus *Dinofelis* represents a line of extinct cats, members of which have been found in Eurasia, Africa, and North America in Pliocene and Pleistocene deposits (figure 3.2). These cats are often referred to as "false" saber-tooths, because their dental morphology falls to some extent between that of the true saber-tooths and the living species, with flattened although reasonably short canines. The animals were also midway in size between a large leopard and a lion, perhaps the equivalent of a jaguar. Little detailed information has been published on the bodily proportions, although it is known that useful material has been collected in South and eastern Africa. What we do know, from specimens collected at the Bolt's Farm locality near Johannesburg in South Africa, is that the forelimb had a relatively short forearm, rather like that seen in forest-dwelling species such as the jaguar or leopard, while the rear limb appears to have been relatively gracile. This would indicate that the animal was not a fast runner, and that it probably had a stronger front limb than modern felids in relation to its overall body size, perhaps able to hold and subdue prey animals but not to chase them very easily.

Species formerly referred to the genus *Therailurus* are now considered to belong to *Dinofelis*, and a poorly known species, *Felis cristata*, from the

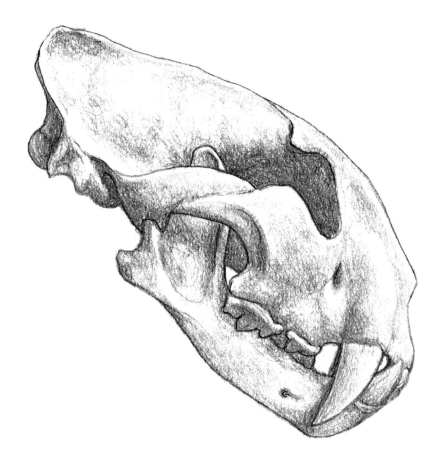

FIGURE 3.2 *Skull of a Chinese* Dinofelis
This beautifully preserved skull comes from Ruscinian-Villafranchian deposits at Shansi in China. It is about the size of *Dinofelis barlowi* from South Africa but smaller than the *D. abeli* or *D. diastemata* type specimens. The mandible has been restored here based on other specimens of *Dinofelis*. A reconstruction of this animal is shown on the cover of the book.

Siwaliks of India and Pakistan, is now also placed in the genus. The earliest specimens have been placed in the species *D. abeli* (China), *D. diastemata* (Europe, plate 3), and *D. paleoonca* (North America). Several good specimens from the South African Mio-Pliocene site of Langebaanweg have been placed in the European species, although they may belong in the African taxon *D. barlowi*, best known from three skeletons found in a pit at Bolt's Farm near Johannesburg (figure 3.3). If all the latter specimens do indeed belong to a single species, then it seems to have been quite sexually dimorphic in regard to size.

The species *Dinofelis piveteaui* is largely known from the South African site of Kromdraai A, where a particularly fine skull was found in deposits dating to around 1.5 Ma ago. The animal shows the most specialized dentition of all the members of the genus, with very flattened although only moderately elongated canines, and with very sectorial cheek teeth lacking some of the robust features seen in the dentition of lions. Unfortunately, very little else is known about this species, although recent discoveries in eastern Africa suggest that the same species may have been present there at about the same time.

Tribe Homotheriini

MACHAIRODUS. *Machairodus* is a genus of large, saber-toothed cat, ranging up to the size of a lion, with elongated upper canines and cheek teeth adapted for slicing meat. These features are discussed further in the next chapter, but the canines in particular are a fine example of the basis for the essential division of the Felidae into saber-toothed and conical-toothed forms.

The genus *Machairodus* first appears in Eurasia in middle Miocene deposits of perhaps 15.0 Ma, and may be recorded as late as 2.0 Ma ago in Tunisian deposits where material referred to the species *Machairodus africanus* has been found. In Europe at around 10.0 Ma a species known as *M. aphanistus* (figure 3.4) is commonly found, and the same species may be present further east in Asia. Recent discoveries at the Spanish Miocene site of Cerro Batallones near Madrid are now giving a fuller picture based on almost complete material (figure 3.5). This species is possibly the same as that known in North America as *Nimravides catacopis*, good samples of which are known from Hemphilian (early Pliocene) age deposits in Kansas, Texas, and Florida (figure 3.6).

Various other species have been named in the literature at one time or another, but at present it remains unclear precisely how many of these will prove to be valid. However, even though the number of species in the genus is not well defined, there appear to be two grades (a term meaning a stage or degree of evolution that may not coincide exactly with a formal, systematic classification). The more primitive grade is exemplified in Europe by *Machairodus aphanistus* and in North America by species of *Nimravides*. The more evolved grade, which includes the Eurasian *M. giganteus* (best known from Pikermi and Samos in Greece) and the very similar American *M. coloradensis*, shows further development of the machairodontine traits in the skull and dentition, such as more-flattened canines and bladelike cheek teeth, a reduced coronoid process of the mandible, and enlarged mastoid processes. In the skeleton, the elongation of the forelimbs is the most obvious development in the latter group, and may be seen in the sequential reconstruction of *M. coloradensis* depicted in figure 3.7. A reconstruction of the head of the Chinese specimen of *M. giganteus* is shown in figure 3.8. In figure 3.9 we emphasize the very real differences between the appearance of this animal and a modern lion; and in figure 3.10 we illustrate a small and rather enigmatic specimen of an undetermined species of *Machairodus* from China.

Crenulations (fine serrations) on the edges of the unworn teeth are characteristic. In most specimens these are commonly seen on the sharp

FIGURE 3.3 Dinofelis barlowi *with a baboon*

The scarce information currently available about the postcranial skeleton of African species of *Dinofelis* points to an animal with the general build of a jaguar. However, the reduction of the metapodia is less than that seen in the American species, and suggestions that *Dinofelis* may have been ambulatory appear exaggerated.

The illustration shows *D. barlowi* attempting to capture a baboon at the South African site of Bolt's Farm. Large male baboons can be extremely difficult for leopards today to overcome, but the extra size and strength of *Dinofelis* may have made it something of a specialist in killing large primates, including our own near relatives and ancestors. At Bolt's Farm the skeletons of three individuals of *Dinofelis* were associated with those of several baboons, and it seems likely that both species had been caught in some kind of natural trap. The presence of coprolites (fossilized droppings) shows that the animals remained alive for some time in the trap before dying, although who remained alive longest is not clear.

FIGURE 3.4 *Study of* Machairodus aphanistus *drinking*
The complete remains of several individuals of this species have recently been found for the first time in the Spanish site of Cerro Batallones. They reveal the rather tigerlike proportions of this animal, most similar to those of the American taxon *Nimravides catacopis*. The dorsal margin of the skull was less straight in profile than in later species of *Machairodus*, giving the animal a more feline appearance, although it would have seemed very narrow to a modern observer when viewed face-on.

front and rear edges of the elongated upper canines; they are present on all the teeth in the fresh and unworn state, but they disappear rapidly as the teeth wear. When examined closely, the unworn crenulations appear almost as beads of enamel set along the edge. Whether they had any significance as adaptive features producing a keener cutting edge is open to serious question. The evidence of wear on the cheek teeth, the ones used for slicing meat, suggests that any such advantage was soon lost during the early life of the animal. At that point the usual phenomenon of a sharp edge preserved by the differential wear of dentine and enamel ensured that the teeth remained capable of cutting despite attrition. But the crenulations appear to have survived longer on the upper canines, and may have provided some advantage to the animal.

The various species of *Machairodus* show an evolution over time toward the more extreme specialization seen in the genus *Homotherium*, to which it is believed to have been ancestral; the Russian paleontologist Marina Sotnikova has suggested that a new species, *Machairodus kurteni* from

FIGURE 3.5 *Life appearance of* Machairodus aphanistus *leaping*
This reconstruction is based on newly discovered material of almost complete
skeletons from the Spanish Miocene locality of Cerro Batallones near Madrid,
excavated by Jorge Morales. The completeness of the Cerro Batallones material
allows us to infer that this species may have had particularly good jumping abilities.

Kalmakpai in Kazakhstan, may be a likely candidate for such ancestry. But, overall, the genus retains more primitive, generalized features. These are most obviously seen in the dentition, where the jaw of *Machairodus* retains a greater number of teeth, and also in some of the bodily proportions as summarized in the figure texts.

HOMOTHERIUM. Lion-sized saber-toothed cats are known throughout Europe and Asia from perhaps 3.0 Ma to 0.5 Ma ago, and the fossils give a rather full picture of morphology (figure 3.11). A number of different species have been proposed for the Eurasian material, such as *H. nestianus*, *H. sainzelli*, *H. crenatidens*, *H. nihowanensis*, and *H. ultimum*, largely on the basis of the size and curvature of the upper canine and perhaps body size. Some of the size differences can be impressive: the length of the skull base of the fine specimen from Perrier in France is 302 mm while that of an individual from Zoukoudian in China is 234 mm; but in view of the range in size and proportions that may be seen in extant cats, it is likely that only one species was present to which the name *H. latidens* may be given (figure 3.12). Members of the genus are also known from Africa, variously referred

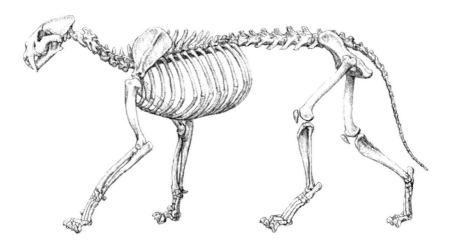

FIGURE 3.6 *Skeleton of* Nimravides catacopis
This illustration, based on very complete remains from various Hemphilian (Miocene) deposits in Kansas, shows the very felinelike proportions of the cat, which lacked the skeletal specializations of later saber-tooths.

Skulls and postcranial bones of this species are abundant in early Hemphilian sites of North America. An associated skeleton from Kansas shows that the proportions of this cat are rather like those of its likely ancestor, *Pseudaelurus*, enlarged to the size of a lion. In life this animal must have operated very much like a large puma, although its great size would probably have prevented it from climbing as efficiently.

Reconstructed height at shoulder 100 cm.

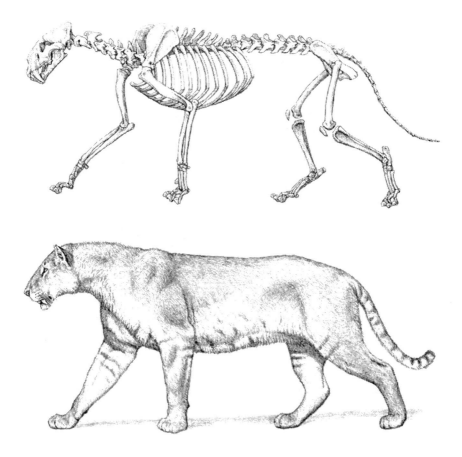

FIGURE 3.7 *Skeleton and life appearance of* Machairodus coloradensis

This specimen is based on material of Hemphilian (late Miocene) age in North America. The proportions of *Machairodus coloradensis* are to some extent intermediate between those of the earlier *Nimravides catacopis* and the later *Homotherium*. There is some elongation of the radius in the forearm compared with *Nimravides*, and the lumbar region of the vertebral column has shortened to resemble that of the living lions and tigers (although remaining longer than in the case of *Homotherium*). Despite the machairodontine adaptations of its head and neck, *M. coloradensis* must have looked very much more like a pantherine cat than *Nimravides* ever could.

Associated limb bones of *M. giganteus* from Pikermi show very similar proportions in the Eurasian form, adding to the observations on the similarities of dental and cranial material made by Gérard de Beaumont.

Reconstructed height at shoulder 120 cm.

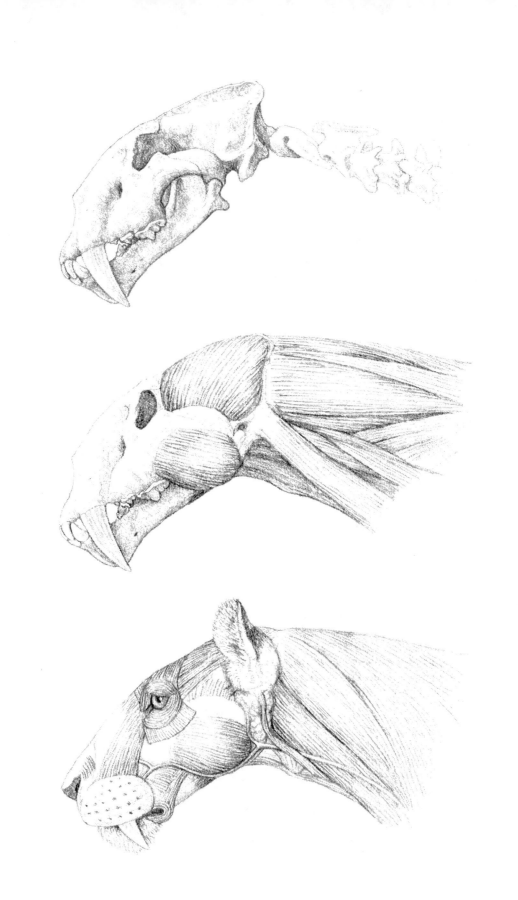

to *H. ethiopicum* and *H. hadarensis,* and North America, where specimens of latest Pliocene to upper Pleistocene age from Alaska to Texas have been referred to *H. serum.*

It is difficult to accept that the known African specimens differ significantly from those of Europe, although the situation for the American material is more difficult to resolve. In the latter case further confusion has been caused at times by the referral of material that more correctly belongs in *Homotherium* to the genus *Dinobastis,* and by the fact that the latter genus has (at least in part) been confused with a third, *Ischyrosmilus.* Part of the material identified as *Ischyrosmilus* seems to be more properly identified as *Homotherium,* but other specimens are more correctly placed in the genus *Smilodon* (see below). Like its probable ancestor among members of the genus *Machairodus, Homotherium* has teeth that display crenulations.

Although *Homotherium* is widely known, many of the occurrences are of fragmentary material. In Europe, a particularly fine skeleton is known from the French site of Senèze in the Auvergne, and skulls and some postcranial remains of at least three individuals have been found at the Spanish site of Incarcal, which may represent the remains of a natural carnivore trap. In North America *Homotherium* is widely distributed but rare and usually represented by isolated bones and teeth. The most striking exception is at the late Pleistocene locality of Friesenhahn Cave in Texas (figure 3.13), where an articulated skeleton was found together with associated remains of more than 30 other individuals of all ages, including cubs. The presence of numerous milk teeth of more than 70 young mammoths in the same deposit has led to interpretations of specialized predation on these animals by *Homotherium,* a point to which we return in chapters 4 and 5.

FIGURE 3.8 *Sequence of reconstruction of the head of* Machairodus giganteus
The best-preserved skulls of *Machairodus giganteus* come from China, and some of them, kept in the Frick Collection at the American Museum of Natural History in New York, still await publication. Formerly known as *M. palanderi,* these specimens were united with the European species by the Swiss paleontologist Gérard de Beaumont.

We can see several differences in comparison with the earlier *Machairodus aphanistus.* The upper canines are thinner, and the lower ones have become relatively smaller. The profile of the roof of the skull is straighter, and the mastoid process of the skull behind and below the ear opening has become more developed. The mandible has a smaller coronoid process (the region to which the temporalis muscle from the skull attaches to produce much of the power for closing the jaw and causing the teeth to bite). The incisor teeth have also developed, becoming a much more significant component of the dental equipment.

Although the head of the living animal as depicted here is reminiscent of that of a lion or a tiger, in reality it would have been longer and much narrower than in either of the two modern cats, as may be seen in figure 3.9.

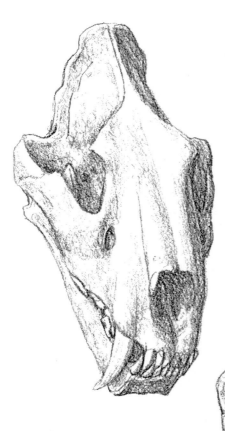

FIGURE 3.9 *Skulls and heads of a lion* (left) *and* Machairodus giganteus

The different proportions of the heads of lions and Miocene machairodonts can be somewhat masked in side view—not least because the long hairs of the chin in lions (and tigers) mimic the angular shape of the mandibular flange in the machairodonts, and because large pantherines may have developed sagittal crests and comparatively long muzzles.

In a more frontal view the differences are more striking, even if we show *Machairodus* with a lionlike, only slightly patterned face. The narrow zygomatic arches (see figure 1.2), the longer muzzle, and the relatively smaller eyes of the machairodont would have been readily apparent in the living animal.

The skeleton of *Homotherium* has many peculiarities when compared with living cats, and even with other fossil forms. Although large and undoubtedly powerful, it was comparatively slender with strikingly elongated forelimbs and a short tail. Interpretations of the lifestyle of this and the other saber-toothed cats have been many and varied, as may be seen in chapter 4, but the appearance of *Homotherium* suggests something unique among the cats.

Tribe Smilodontini

PARAMACHAIRODUS. Opinions on the taxonomic position and relationships of this genus are rather divided. After a long and confusing history of

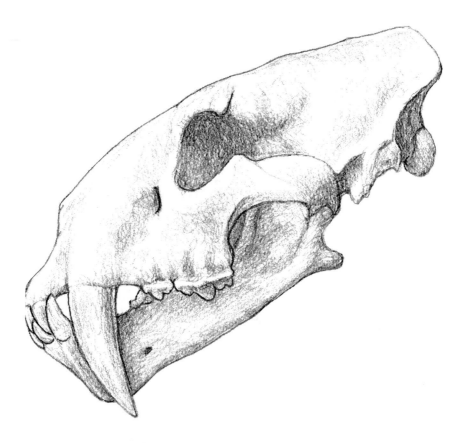

FIGURE 3.10 *Skull of a small* Machairodus *from China*
This specimen comes from Pliocene-age deposits at Fan-Tsun in China. The skull is very small for *Machairodus*: a skull of *M. giganteus* from Shansi is 37% larger, with a basal length of 315 mm against one of 229 mm in the smaller individual. The mandible and part of the zygomatic arch have been restored for the purpose of illustration.

discussion by various authors, many of the species formerly placed in this genus have now been reassigned to the genus *Pontosmilus* within the tribe Metailurini, leaving only *Paramachairodus ogygia* of Vallesian and earliest Turolian age and the later *P. orientalis*. Until recently *P. ogygia* was poorly known, but a remarkable series of specimens have now been recovered from the Spanish Miocene locality of Cerro Batallones near Madrid, with an age of perhaps 9.0 Ma; almost-complete skulls and skeletons are now known for the first time, and these allow a better assessment of the species' morphology and relationships (figures 3.14 and 3.15).

MEGANTEREON CULTRIDENS. Members of the genus *Megantereon* have been found in Africa, Eurasia, and North America (plate 4). Whether all belong to the same species remains unclear, since some of the American specimens are fragmentary, but the African and Eurasian ones certainly appear to be conspecific. The origins of the species are also uncertain, but it was present in both Africa and Eurasia by 3.0 Ma ago, and in North America perhaps shortly after that time. It has not so far been recorded in British deposits, and its western European distribution seems to have been generally confined to more southerly parts.

Megantereon cultridens has never been recorded in abundance, and at many localities it is known from single individuals—usually by pieces of mandible or the characteristic upper canine, and rarely by parts of the skeleton (figure 3.16). But it is present at a fairly large number of European sites before its last appearance in that continent at the German locality of Untermassfeld at around 0.9 Ma ago. The most remarkable find is the complete skeleton from Senèze in the French Auvergne, which is now kept in the Natural History Museum in Basel. Much of our reconstruction and discussion of the morphology and possible behavior of the animal (figure 3.17) is based on this skeleton, and upon the preliminary description of it published by Samuel Schaub.

The Senèze skeleton is from an animal the size of a large leopard, with a shoulder height of around 70 cm. It was a powerfully built cat, with especially massive forelimbs and claws the size of a lion, and its bodily proportions suggest that it was able to bring down and hold reasonably large prey (figures 3.18 and 3.19 and plate 5). Some isolated limb bones from the South African site of Elandsfontein, of broadly similar age to Senèze, are even shorter in the distal limb elements, such as the tibia, and serve to confirm the powerful structure of the animal. Yet its means of killing its prey remain obscure. Unlike modern larger cats, which have moderately long, conical upper and lower canines well suited to stabbing, *Megantereon* had long, flattened upper canines and relatively small lower ones. The upper canines appear likely to have been easily broken if sunk into a struggling animal, and yet the length and strength of the neck and the wide gape of the lower jaw have led to suggestions that the animal struck with its mouth

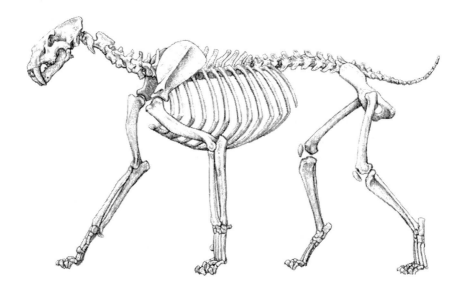

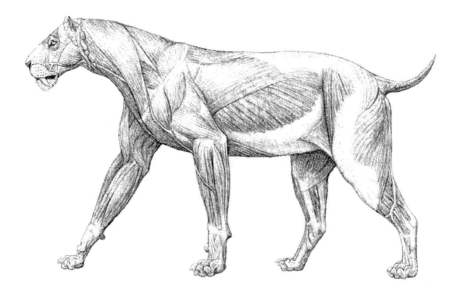

open, inflicting a deep wound. Many of the arguments about the killing technique employed by the machairodont cats lack practicality and common sense, and we shall discuss this problem further in the next chapter when we deal with anatomy and action.

SMILODON. Cats of the genus *Smilodon* are known from North and South America, but they have never been found in Eurasia. Members of the genus are among the most recent of the saber-toothed species, having become extinct in North and South America toward the end of the last glaciation around 10,000 years ago. These cats are among the best known of the fossil carnivores thanks to the enormous numbers of them found in the asphalt deposits of Rancho La Brea in Los Angeles, California: the huge sample from there consists of approximately 162,000 bones of at least 1,200 individuals, from a total of perhaps one million bones that have been collected at the site (figures 3.20 and 3.21).

Despite a complex history of names, three species are now generally recognized. The earliest is *Smilodon gracilis*, known mainly from the eastern part of the United States of America, aged between perhaps 2.5 and 0.5 Ma. It is the smallest of the species and is thought to be most closely related to *Megantereon*, its likely ancestor. The largest, *S. populator* (figure 3.22), was of lion size, with enormously developed upper canines that protruded well below the mandible: the total length of large tooth specimens approaches 28 cm, perhaps 17 cm of which would have protruded from the upper jaw (figure 3.23). *Smilodon populator* is a species found in the eastern part of South America; it seems to have evolved in isolation there following dispersion of the ancestral *S. gracilis* from the north about 1.0 Ma ago (plate 6). The third species, *S. fatalis*, primarily known from the later Pleistocene

....

FIGURE 3.11 Homotherium latidens *skeleton and a sequential reconstruction*
The limb proportions of *Homotherium latidens* are rather difficult to understand from a functional point of view, and yet the trend seen in the European Villafranchian form is carried to extremes in the late Pleistocene representatives from Friesenhahn Cave in North America. Some features of the hind limb in particular were once taken to indicate a plantigrade form of locomotion, with the foot placed fully on the ground, but it is now recognized that these interpretations are incorrect.

In comparison with later species of *Machairodus*, the specimen of *Homotherium latidens* from the French site of Senèze has a shortened lumbar region, an elongated radius, and forelimbs that are longer in proportion to the hind limbs than in "normal" cats. There is also an unusually short calcaneum. The fact that such characters are seen in even more extreme form in the Friesenhahn Cave animals would imply that they had real selective advantage. In general, such limb proportions are associated with a reduced ability to jump and leap, and it has been suggested that *Homotherium latidens* falls into an intermediate position between pantherine cats and hyenas. See the text for a fuller discussion of these points.
Reconstructed height at shoulder 110 cm.

of North America, is generally intermediate in size, but with important differences from the South American species in skull and bodily shape and proportions. *S. fatalis* also invaded South America, where it is known from the Pacific coastal area; the two species in South America were therefore separated in their distributions by the Andes (figures 3.24 and 3.25).

Many studies have been made of the massive Rancho La Brea sample. It provides a detailed insight not only into the size and natural variation of the animal, but also into some of the problems that afflicted it. At least 5,000 of the bones, drawn from every part of the skeleton, show some kind of pathological condition, ranging from developmental abnormalities and dental disease to wounds and stress injuries. Many of the latter appear to have resulted from overextension of limbs leading to the tearing of ligament and muscle attachments, and they show degenerative osteoarthritis. Some bones, especially from the shoulder and spine, show signs of healed wounds that suggest fighting, in some cases with other cats.

FIGURE 3.12 *Life appearance of* Homotherium latidens
Our reconstruction is based on one of the best and most complete skeletons of *Homotherium* from Senèze in France, which is kept at the Université Claude-Bernard in Lyon and was described in detail by Rolland Ballesio.

FIGURE 3.13 *Reconstruction of* Homotherium serum *from Friesenhahn Cave*
The American members of the genus have been referred to this species.

FIGURE 3.14 *Reconstruction of the skull and living head of* Paramachairodus ogygia
These sketches show the narrow head and long muzzle that probably made the
animal look superficially similar to the living clouded leopard.

SUBFAMILY FELINAE

Tribe Felini

We have chosen to place the living and fossil conical-toothed cats in a single subfamily and tribe, but it is worth underlining the point that this is not the view of all authorities. Some would advocate splitting the living species into a number of distinct subfamilies, such as the Pantherinae to include the genera *Panthera* and *Lynx*, and the Felinae to accommodate the genus *Felis*, with the position of the cheetah unresolved.

PANTHERA. The genus *Panthera* is generally considered to contain the living lion, leopard, tiger, and jaguar, with the possible addition of the snow leopard, and the fossil species *Panthera gombaszoegensis* and *"Panthera" schaubi*. For most authorities, the major linking feature is taken to be the presence of an elastic ligament in the hyoid apparatus of the throat, a system of small bones that provides support and anchorage for the voice box and tongue (figure 3.26). This ligament is generally thought to permit movement and the production of the characteristic roar of these animals, which contrasts with the variety of quieter calls in other species. If the feature has indeed been derived from an earlier, ligament-free condition, then it would link the species that exhibit the character. However, the correlation between roaring and the presence of a hyoid ligament, and thus the

FIGURE 3.15 *Life reconstruction of* Paramachairodus ogygia

The first reasonably complete remains of this cat come from the Spanish site of Cerro Batallones, and reveal that the leopard-sized creature was built much like the ancestral cat *Pseudaelurus*, with a versatile, supple body and powerful forelimbs. When compared with a leopard, *Paramachairodus* is now seen to have had longer and more gracile hind limbs and more robust forelimbs. These traits, combined with its body size, point to an agile climber and a hunter of relatively large prey, although the head was rather small and narrower than in the leopard.

Reconstructed height at shoulder 58 cm.

taxonomic significance of the character, has recently been questioned by Gustav Peters and M. H. Hast.

The pantherine cats are often considered to have had the most recent radiation into distinct species, with molecular biologists such as Stephen O'Brien and his coworkers suggesting that this has only taken place within the last two million years or so. However, this argument is not supported by the current fossil evidence, and there is no particular reason to assume that the great cats, as they are often known, are the most recently evolved of the living species. The earliest history of the genus *Panthera* is simply not well known, and while molecular biology has a clear part to play in assessing

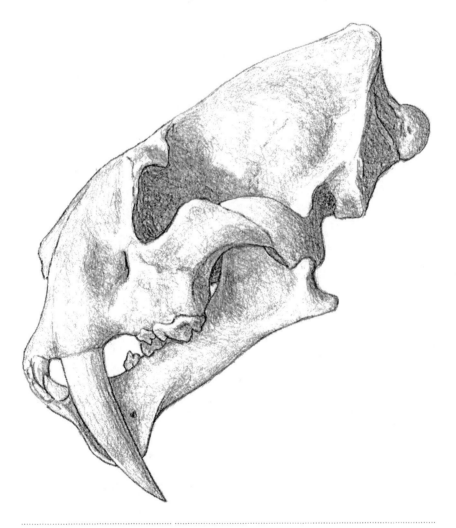

FIGURE 3.16 Megantereon cultridens *skull*
This reconstruction is based on a skull and mandible from Nihowan in China described by Pierre Teilhard de Chardin and Jean Piveteau, and an upper canine from Shansi.

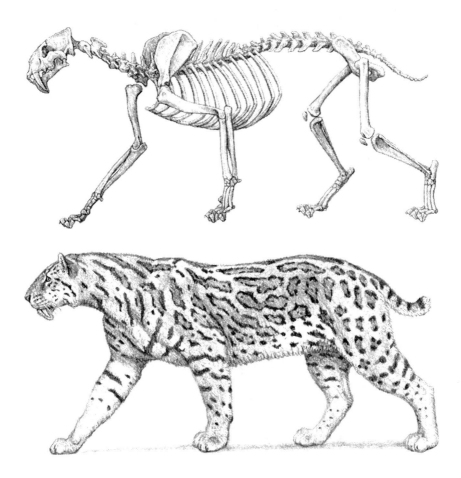

FIGURE 3.17 *Skeleton and reconstruction of* Megantereon cultridens

Megantereon was a powerfully built cat, but its limb proportions are still within the range observed for living species. Robust individuals of the jaguar often have relatively shorter distal limb segments than the Senèze individual, while shortening in *Smilodon* may be more extreme than in either. The main difference between the postcranial morphology of *Megantereon* and living cats lies in the lengthened neck and shortened lumbar region of the spine.

Given its size and build, *Megantereon* is likely to have been quite as adept at climbing trees as leopards and jaguars, although its shortened and stiffened lumbar region may have reduced maneuverability somewhat. But that stiffening would have increased the strength available for overcoming prey, while the length of the neck was most probably related to the method of biting with its enormous canines, as discussed further in the text.

Reconstructed height at shoulder 72 cm.

relationships and patterns of branching it is by no means an infallible guide to the timing of such events.

We begin this section with one of the most enigmatic of all the fossil cats, *"Panthera" schaubi*. The taxon was originally found at the French site of Saint-Vallier in the Rhône Valley, in deposits believed to date to around 2.1 Ma ago. The site produced an extremely rich assemblage, but only three specimens of this particular cat: a skull and two mandibles. Although the specimens are in good condition, the identity of this cat has remained something of a mystery. The skull is of an animal about the size of a small leopard or a very large lynx—but it is too short-faced to be a leopard, which in any event does not appear to have dispersed into Europe for another 1.0 Ma or so, and it is clearly different from the Plio-Pleistocene *Lynx issiodorensis*, which is well represented at the site. Its allocation to the genus *Panthera* has been questioned by some, and a new genus, *Viretailurus*, has been proposed. Other suggestions have included a possible link with the American puma.

The only other known possible occurrences are at the Spanish sites of Venta Micena, in deposits dating to around 1.2 Ma ago, and at La Puebla de Valverde, in deposits dating to around 2.2. Ma ago. But both these identifications are uncertain.

Panthera onca, the jaguar, is the largest cat, and indeed the largest member of the Carnivora, in Central and South America (figure 3.27). Males

FIGURE 3.18 Megantereon cultridens *stalking*
This scene and the following scene are set in wooded territory of the French Massif Central during the earliest part of the Pleistocene. In this first scene the cat is shown stalking.

FIGURE 3.19 Megantereon cultridens *with two deer*
In summer the cats would have hunted near the streams in the wooded bottoms of the valleys, perhaps waiting until a suitable victim appeared to quench its thirst. Among the favored targets are likely to have been the young of larger ungulates, as is seen today in eastern Africa. The only difference in Europe is likely to have been the presence of various deer in place of the numerous African antelopes. In this scene *Megantereon* is shown preparing to launch itself at a mother and calf of the cervid genus *Eucladoceros*. In chapter 5 we discuss the reasons why male deer would have been a more dangerous prey for this cat.

may weight up to 120 kg, and females up to 90 kg. It is a compact, strongly built animal, not unlike the leopard in superficial appearance but generally larger and more stocky and with a somewhat different coat pattern. Both have the same dark spots, but those of the jaguar are larger and less numerous (figure 3.28). As with the leopard, all-black animals are relatively common, and fall within the normal range of individual variation in coat coloring.

Jaguars are recorded in North America from earliest Pleistocene deposits, so that the fossil history of the species spans approximately 1.5 Ma. Although now confined to the southern half of the Americas, it ranged over both northern and southern parts during the Pleistocene, at times up to Washington and Nebraska, and was often larger than the modern form. It is most abundantly represented as a fossil in areas where the lion was scarce, especially in Florida, Texas, and Tennessee. It is probable that it was most closely related to the European species *P. gombaszoegensis*, often referred to as the European jaguar.

Overall, its main prey may be said to consist of capybara, tapir, marsh deer, peccary, armadillo, and paca, but the jaguar is a strong swimmer and may eat fish, turtles, and even smaller alligators. In Venezuela jaguars kill domestic animals including cattle, and although they usually take juveniles they have been known to kill bulls weighing up to 500 kg. In Brazil they are able to deal with the aggressive white-lipped peccary, which pumas in the same region generally avoid. They are largely solitary cats, breeding seasonally in the temperate parts of their range but all year round in the tropical regions where climatic changes are less marked.

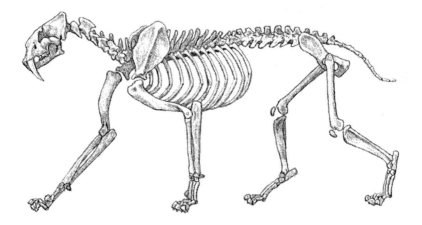

FIGURE 3.20 *Skeleton and life appearance of* Smilodon fatalis
Reconstructed height at shoulder 100 cm.

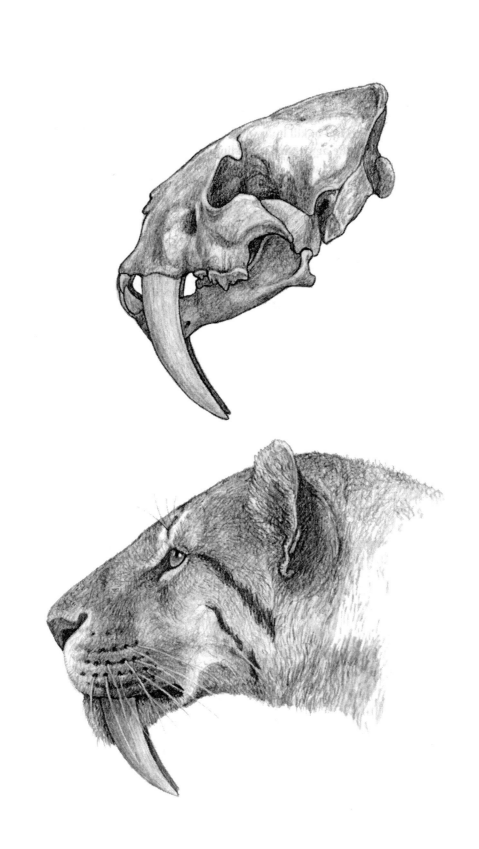

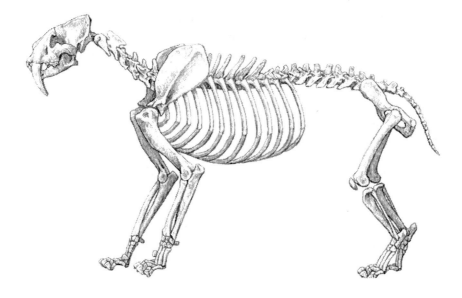

FIGURE 3.22 *Skeleton of* Smilodon populator

The proportions of the limbs and vertebral column in *Smilodon* resemble those of *Homotherium* (figure 3.11) in the shortened lumbar zone and powerful forelimbs. But while the skeleton of *Homotherium* is no more robust than that of a living lion, *Smilodon populator* had very broad limb bones and a marked shortening of the distal limb elements. In *Homotherium serum*, for example, the radius was typically 91% of the length of the humerus, while in *Smilodon populator* it was only 73%. *Smilodon* also had a longer calcaneum and may have been a better jumper and leaper than *Homotherium* in spite of its heaviness, since the increased length would have increased leverage at the ankle joint. One other point of similarity between the two genera may be mentioned: both tend to have a high scapula, although the functional significance of that is not clear to us.

Reconstructed height at shoulder 120 cm.

FIGURE 3.21 Smilodon fatalis *skull and head*

Based on skull 2001-2 from Rancho La Brea. The development of the upper canines in *Smilodon* can be compared only with that seen in *Barbourofelis* and *Thylacosmilus*. Unlike those two, however, *Smilodon* lacks a developed mandibular flange.

Smilodon differs from *Megantereon* in its larger size, longer canines, lack of flange, and reduction of the premolars. The occipital region of the skull is also more inclined in *Smilodon*, although South American specimens show a secondary elevation of the occipital plane. The late Pleistocene animals from South America also differ from the typical *Smilodon* of Rancho La Brea (the kind depicted here) in having a straighter profile to the top of the skull and a greater width across the zygomatic arches at the side of the head.

The living jaguar shows an interesting pattern of size variation over its geographic range: the smallest specimens are found in equatorial regions, and the size gradually increases toward the northern and southern limits of the range. This pattern of size variation, known as a cline, is most probably linked to a combination of climatic and habitat factors, since it is advantageous to be of larger body size in colder conditions in order to reduce heat loss, as well as in more open terrain where larger prey tend to be encountered. Thus the largest living jaguars are those living in less-forested habitats. Size clines are seen in many other members of the Carnivora, although few have such a double pattern of change in two directions. Of those that do, the brown bear has one of the most interesting, with reductions in size to both the east and the west of the Beringian area where Siberia and Alaska almost meet and the largest bears now occur. It was across Beringia that the American and Eurasian faunas, no doubt including the jaguars, inter-

FIGURE 3.23 *Life reconstruction of* Smilodon populator
Smilodon specimens from the Lujanian or late Pleistocene of South America were enormously powerful animals. Compared with the La Brea specimens they exhibit shortened distal limb segments and very robust metapodia. The forelimbs are extremely strong, making the hind limbs look weak in comparison. These features may mark adaptations for dealing with the slow-moving "megafauna" of the continent, animals such as toxodonts and megatheres that were themselves enormous and ponderous. It has even been suggested that *Smilodon* may have contributed significantly to the extinction of these large, indigenous herbivores.

changed during glacial times of the past million or so years when the sea was lowered, and the clines in bear sizes point to the former existence of a common population.

In addition to the size cline in living jaguars there is also evidence of an overall trend toward smaller size over time in the lineage, with modern animals considerably smaller than their Pleistocene forebears. More recent jaguars also exhibit shortening of the limbs, especially of the metapodia (the bones in the equivalent of our palm or the sole of our foot), which Björn Kurtén suggested may indicate an increased specialization on life in forests and broken country. It is possible, in view of the tendency toward mutual exclusion with the lion in fossil deposits, that the jaguar was driven from its earlier range in more open terrain by the later appearance of the lion, and had to modify its habits in the process, with eventual selection for a more appropriate body plan.

The extinct European jaguar, *Panthera gombaszoegensis* (figure 3.29), is first recorded at Olivola in Italy at around 1.6 Ma ago, and is known in some numbers from other Italian Villafranchian localities under the synonym of *P. toscana.* It is then present until well into the middle Pleistocene at sites such as Westbury in England, Mosbach in Germany, Atapuerca in Spain, and L'Escale in southern France. It seems to have been a large and robust animal, bigger than the extant New World jaguar, *P. onca*, and broadly sim-

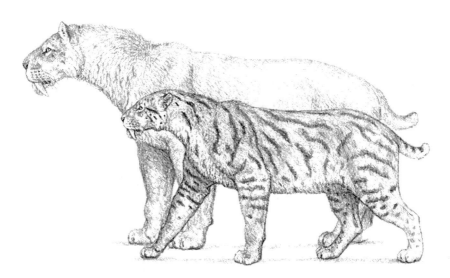

FIGURE 3.24 *Size comparison between* Smilodon populator (*left*) *and* Smilodon fatalis
The smaller animal in this drawing is based on an incomplete specimen of *Smilodon fatalis* from Florida, which is slightly below the average size of the Rancho La Brea sample of *S. fatalis*. The larger one is based on a complete skeleton of *S. populator* from Argentina, a large one but not the largest. The difference in size is striking.

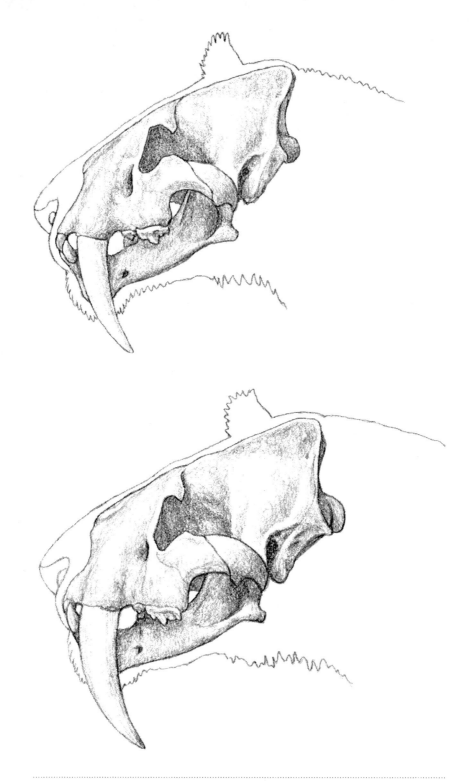

FIGURE 3.25 *Skull and head differences between* Smilodon fatalis (top) *and* Smilodon populator

It is clear from the drawings that the skull of the North American *Smilodon fatalis* has a more convex profile to the top of the head (somewhat reminiscent of *Megantereon*) than its South American relative.

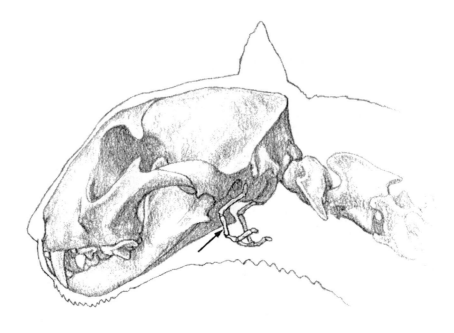

FIGURE 3.26 *Hyoid apparatus*

The hyoid is made up of several bones. The epihyal (indicated by an arrow in the drawing) is not ossified in the pantherine cats. This trait has been thought to be the most important determinant of roaring ability, but recent research points to other factors that may be of equal or even greater significance.

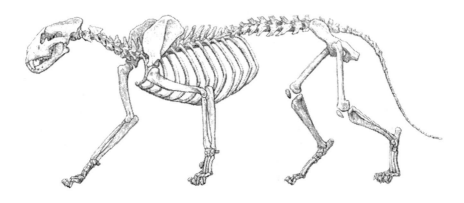

FIGURE 3.27 *Skeleton of the jaguar,* Panthera onca

The living jaguar is a very robust animal, perhaps the most robust of all felids with the exception of *Smilodon* and some individuals of *Megantereon*. The skull is proportionally large, with prominent crests and a relatively long muzzle. The forearms are extremely powerful for the size of the animal, and they help it take large prey while also improving its climbing ability; in fact, the jaguar is the heaviest cat that climbs regularly and proficiently.

Reconstructed height at shoulder 70 cm.

FIGURE 3.28 *Life appearance of the jaguar*
In spite of the jaguar's stocky build and large head, the similarity between it and the leopard is striking and may indicate a very close relationship. Of all the large cats, only the jaguar, leopard, and lion share this particular rosetted coat pattern, and recent research by Gustav Peters shows that they share something else: a structured call sequence, or "roaring proper," not present in the tiger or snow leopard.

ilar in size to the fossil specimens of that region. It is particularly well represented at the middle Pleistocene site of Westbury-sub-Mendip, just to the south of Bristol.

Panthera leo, the lion, is presently found in much of the open country of Africa south of the equator, and as a small relict population in the Gir Forest Reserve of northwestern India estimated at under 300 individuals in a 1990 census. In historic times it was found throughout Africa, Arabia, Greece, and much of northern India, and was last reported in Iran as recently as the 1940s. Even the relict Indian population was considerably more widely distributed in the first half of the last century. During the upper Pleistocene it ranged across Eurasia and eventually into the Americas (figures 3.30 and 3.31), where it finally reached as far south as Peru. The American variety (plate 7) has often been referred to a separate species, *P. atrox,* and recent work by Sandra Herrington suggests that confusion over the question of whether or not it is actually a lion has perhaps been caused by the presence of some tigers in the northernmost part of North America during the past hundred thousand years or so.

The earliest record of the lion is at Laetoli in Tanzania at perhaps 3.5 Ma ago. Bizarre as it may seem for so large an animal, we simply have no fossil record for it before that, or even for a likely ancestor. It first appears in Europe at the Italian site of Isernia, around 0.7 Ma ago, and from then on it becomes a reasonably common member of the European fauna. Several deposits from the last glacial and interglacial cycle, including

Gailenreuth (Germany) and Wierzchowskiej Gornej (Poland), have produced moderately large numbers of individuals.

Numerous taxonomic schemes have been proposed to account for the variations seen in the European specimens: some authors have even suggested the presence of tigers rather than lions, but this is without substantiation. A major source of confusion seems to have been the importance attached to size. Extant African lions exhibit considerable geographic variation in size, with those from the southern part of the continent being larger (on average) than those from eastern parts. This pattern is further complicated by the fact that the males in any population are considerably larger than the females. Males in eastern Africa may average about 170 kg, while females average about 120 kg; in southern Africa an average of about 190 kg for males has been recorded, with a maximum of 225 kg. When these two factors—geography and sexual dimorphism—are taken into account, the extent of variation within the European sample may be easily accommodated within a single species. Further, the extent of sexual dimorphism means that the size of an individual specimen can be extremely misleading as a guide to the typical size of lions in a given area at a given time, a point that must be borne in mind for all cats. On average, European lions were larger than extant members of the species, and some were very large indeed—but the overlap in ranges is considerable.

The pride structure of African lion society, together with the existence of nomad male groups, would imply that fossil assemblages of lion bones may be largely composed of the remains of a single sex. Since males are bigger than females, such single-sexed fossil assemblages of larger individuals would have no taxonomic significance. However, they might tell us something about the social behavior of the local population.

The most important prey species for lions today are wildebeest, zebra, buffalo, hartebeest, giraffe, and warthog. Their hunting behavior, discussed more fully in chapters 4 and 5, suggests that horses and larger deer may have been favored as prey in Europe, although the greater size of the Pleistocene lions may mean that some of the larger bovids were also a target.

Panthera tigris, the tiger, has a historical distribution that stretched from the Caspian Sea in the west through India and southeastern Asia (including Indonesia) to northern China and Siberia. The tiger reaches the greatest size among the living cats, although size variation between populations, and between males and females in the same population, is considerable. Whereas a Javan male may weigh up to 140 kg and a male Bengal tiger may weigh up to 260 kg, Siberian males commonly range up to 280 kg, and have even been recorded as reaching 384 kg (figure 3.32).

In contrast to the lion, the tiger is usually considered a solitary animal that seems to prefer more closed vegetation, where it hunts by stalking. Favored prey include sambar deer (the largest southeastern Asian cervid), the smaller axis deer, wild pig, gaur (the largest of the wild cattle), and langur monkeys. Whether solitary hunting is a relatively recent development in the tiger's social and predatory behavior is unclear, but observations of its behavior at bait sites to which it has been lured for study suggest that it may be as potentially sociable as the lion. In all probability we should recognize a degree of flexibility in the behavior of both species, and bear in mind that observations tell us only what is done and not necessarily what may be done. (See chapter 5, where we discuss the patterns of social interaction in the living felids more fully.)

The fossil history of the tiger goes back perhaps 1.5 Ma in essentially the same area as its modern distribution, although Russian paleontologists have recorded it in later Pleistocene deposits of the Caucasus. However, the identity of some of the fossil specimens may be open to question. Some fossils have also been recorded from northern Siberia, far to the north of the historical distribution, raising the question of whether the species could, like the lion, have crossed the Bering land bridge into the Americas at times of lowered sea level. Despite their similarity, the separation of lion and tiger skeletal remains is possible (figure 3.33), and, as we have already pointed out, recent studies by Sandra Herrington of skulls from eastern Beringia (modern-day Alaska) now suggest that both lions and tigers were present there within the past 100,000 years during the last glaciation.

Panthera pardus, the leopard (figure 3.34), is today found in many parts of Africa, the Near and Middle East, much of southern Asia, and the Malaysian archipelago, and it is the most widely distributed of the living larger cats. Prey across this vast range varies greatly, but it includes gazelles, impala, young wildebeest, baboons, porcupines, bushbuck, deer, and wild goats. Leopards are said to be particularly partial to dog, although whether this reflects their response to being hunted by such animals or a true dietary preference is unclear. During the Pleistocene the leopard was also present in much of Europe, but it does not appear to have crossed Beringia into North America. Its wide distribution today and in the past underlines the generalized abilities of the larger cats to cope with a variety of environments when given the chance.

As in the case of the lion, the leopard is first recorded at Laetoli in Tanzania, and its first European appearance is at Vallonnet in the south of France, at around 0.9 Ma. Its subsequent record in Europe is patchy,

FIGURE 3.29 *Skull and head of* Panthera gombaszoegensis
While this animal is widely known from a variety of deposits, no fully complete skulls have been found. Our reconstruction is based on the partial skull from the Mygdonia Basin described by George Koufos. Although the specimen is quite eroded and lacks the zygomatic arches, it still gives an idea of the overall proportions. The missing parts of the skull, and the missing mandible, are shown in lighter color.

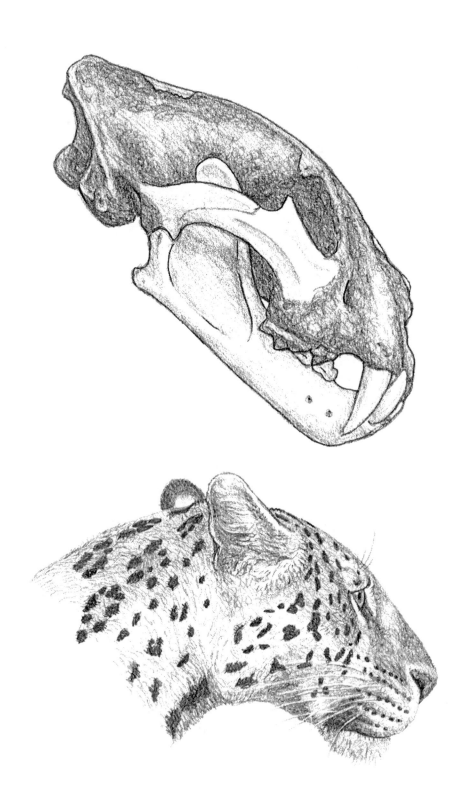

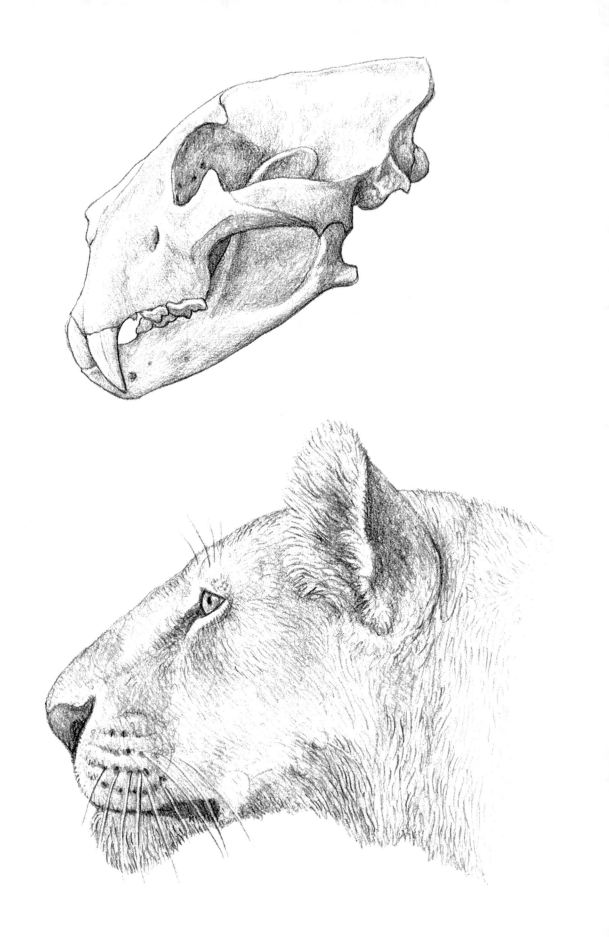

and the only site where it is preserved in any numbers is the Italian last-glaciation cave site of Equi. Modern leopards are by nature largely solitary and secretive, and the fossil record is unlikely to offer a true reflection of their real prevalence.

Like other cats, the leopard is a sexually dimorphic animal that also exhibits considerable geographic size variation, making interpretations of the size of individual fossil specimens rather difficult. Large modern males in eastern Africa may weigh up to 70 kg, while those from the Cape in the south may be half that size. However, all indications are that the European representatives were generally as large as the largest of extant leopards. Such considerations are important in efforts to assess predatory activities. The leopard is a strong animal, able to drag large carcasses (figure 3.35) and to haul them up into trees when faced with the threat of stealing by hyenas (see figures 5.25 and 5.27).

Panthera uncia, the snow leopard (figure 3.36), is a medium-sized cat with a fairly restricted range centering on northern Afghanistan, Pakistan, India, and regions bordering the Himalayas and northward into Mongolia. Its fossil history is sparse; until recently, it was thought to consist of only a few late Pleistocene remains in caves of the Altai mountains on the western borders of Mongolia, based on material recovered up to 100 years ago. However, more recent work based on dated samples from the upper Siwalik deposits of northern Pakistan dated to 1.2–1.4 Ma ago has now shown that the snow leopard was very probably present there at that time.

Panthera uncia is similar in size to the common leopard, with males weighing up to 75 kg, although its magnificent coat and its long, thick tail tend to make it look somewhat larger. The pattern of the coat is broadly similar to that of the leopard and jaguar, but the background color is much lighter. While the size range of prey taken is similar to that of the leopard, local availability dictates what is actually eaten; bharal (blue sheep), markhor (one of the largest of the wild goats, with thick and corkscrewlike horns), and marmots make up much of the diet.

There is some question about the true relationship of the snow leopard, since both skull shape and bodily proportions, together with its relatively short and rounded canine teeth, have suggested to some a closer affinity with the cheetah (figure 3.37). The top of the skull is broad and domed, providing enlarged nasal chambers and air sinuses that may well be linked to life in a colder environment. It has never been heard to roar, and there appear to be some doubts over the exact nature of the hyoid apparatus. For these reasons, the snow leopard has been placed by some authorities in a

FIGURE 3.30 *Skull and head of the American lion,* Panthera atrox
Although often referred to a different species in the literature, it seems likely that the American lion belonged to the same taxon as its African and Eurasian relatives.

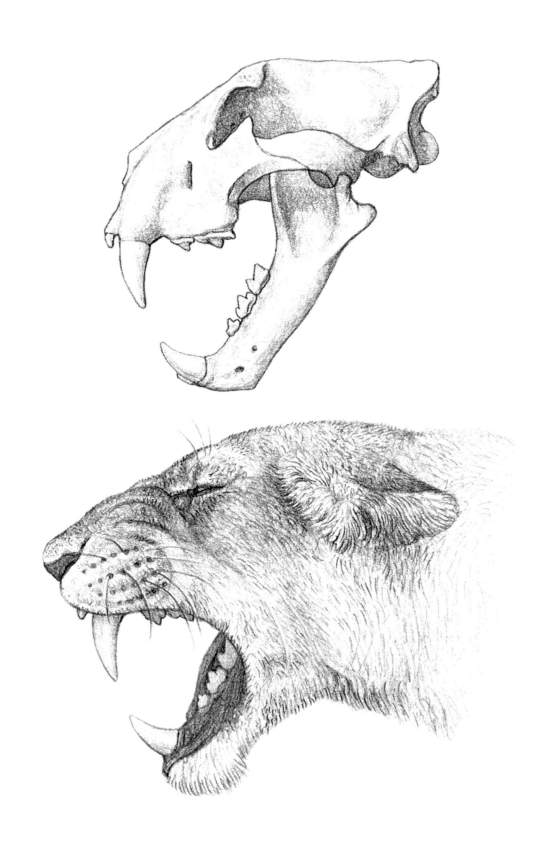

these reasons, the snow leopard has been placed by some authorities in a separate genus, *Uncia.*

The coat has led to the population of this cat falling alarmingly in the wild, since its obvious beauty and high value have meant that the animal is hunted relentlessly. Scarcity of food and the harshness of the terrain in which it exists combine to ensure a solitary lifestyle and hence a low population density, which adds to the danger of extinction once undue pressure is brought to bear.

The paws of the snow leopard have developed quite thick cushions of fur between the pads. These provide insulation to help the feet stay warmer in the winter and cooler in the summer, when hot rock surfaces can be uncomfortable. Such fur may also act as a kind of snowshoe, spreading the weight somewhat and helping the animal to walk across snow without sinking too deeply—a development seen also in the lynxes of the northern regions.

FELIS. The only species of larger cat usually placed in the genus *Felis* is *F. concolor,* variously known as the mountain lion, the puma, or the cougar. There have been suggestions that it should be reclassified and placed in its own separate genus as *Puma concolor,* in view of its distinctiveness; we have no strong opinion on the matter, and therefore we have decided to retain the genus *Felis* for our purpose here. The species is today found over a wide area of the Americas, from British Columbia to Patagonia. Like the leopard, it therefore exhibits the wide tolerances of which larger cats are capable, especially those that operate alone and secretively. It is a moderately large and extremely powerful animal that may reach up to 103 kg in a large male, which means that it is intermediate in average size between the leopard and the jaguar (figures 3.38 and 3.39). It is capable of killing prey up to the size of red deer (or elk, as they are known in North America), although white-tailed deer, mule deer, marsh deer, peccary, and guanaco are among the most common items of the diet. Pumas make frequent use of caves as dens, and seem to show some preference for more wooded terrain.

A marked characteristic of the modern puma is the extent of size variation, which increases significantly to both the north and the south of the

FIGURE 3.31 *Skull and head of a European cave lion,* Panthera leo
The remote ancestors of lions were very likely spotted, leopardlike animals, but the European lions of the Pleistocene were plain-colored, just like their living relatives. This is made clear by the depictions of lions in cave art of the upper Paleolithic, including the paintings in the recently discovered Chauvet cave in the Ardèche valley of southeastern France. These renderings show accurate details, such as the black dots at the base of the whiskers, and it is striking that no maned individuals are portrayed. Did the European male lions lack manes, or are we simply seeing the typical, female-dominated pride structure of the living animal?

FIGURE 3.32 *Life appearance of the tiger,* Panthera tigris

Tigers vary greatly in size and appearance over their modern range. Shown here
are two individuals from Siberia, the largest variety of living tiger.

Reconstructed height at shoulder 95 cm.

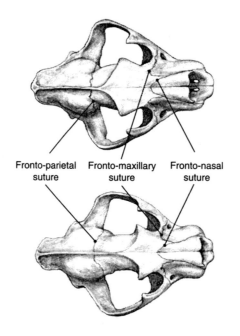

Fronto-parietal
suture

Fronto-maxillary
suture

Fronto-nasal
suture

FIGURE 3.33 *Some distinctions between skulls of lions (top) and tigers*

Although the coat patterning of the lion and the tiger gives them a very different
appearance, their skeletons are very similar in size and overall appearance.
Nevertheless, it is possible to distinguish the two using criteria devised by the
French paleontologist Marcelin Boule around the turn of the century. Shown
here are some of the differences between the skulls when viewed from above; note
especially the fronto-parietal, fronto-maxillary, and fronto-nasal sutures.

equator and once more warns us of the dangers of using size alone as a means of distinguishing between fossil species.

The puma is unknown outside the American continent. Its fossil record there is reasonably good for the past half million or so years, but its earlier history is unknown. It has been suggested that it may bear some closer relationship to the cheetah and to the American cheetah-like cats placed in the genus *Miracinonyx*, which are discussed further below. It is therefore interesting to note that some recent studies of the biomolecular structure of the puma have indeed suggested that its closest living relative may be the cheetah, with a split occurring some time after 3.5 Ma ago.

ACINONYX. The living cheetah, *Acinonyx jubatus*, is historically recorded in Africa, Asia, and the Near East; it is now largely confined to isolated populations in Africa. The oldest record of the species is in eastern and southern Africa from about 3.5–3.0 Ma ago, but a large variety occurs in Europe and Asia from almost the same date. The European form has been referred to a separate species, *A. pardinensis* (figure 3.40), and is known from there as late as 0.5 Ma ago at the German site of Mosbach. The distinction between the two is largely based on size, and it would seem that the African and Eurasian specimens might logically be placed in a single species.

The European cheetah was considerably larger than the modern form, in which the maximum weight is around 60 kg, as illustrated in figure 3.41. Since it had the body proportions of the living cheetah, and since running speed is a reflection of stride length for a given stride frequency, such large animals may also have been capable of running somewhat faster than their living relatives, although greater weight may have countered any advantage of greater size. Whether they *needed* to run faster is less clear. Generally speaking, representatives of a given species often tend to be larger in colder climates, because a bigger body conserves heat better (as we mentioned in the case of the jaguar). The European cheetahs may simply reflect that fact, and the higher potential speed may have been a mere by-product. Greater size would also be of help in subduing prey, so that we can suggest two nonexclusive alternatives to the interpretation of selection for greater speed. The prey of living cheetahs tends to be primarily gazelles, impalas, and springbok, and young zebra may also be taken; prey taken by fossil cheetahs is a matter of inference (figure 3.42), as we discuss further in chapter 5.

The total number of fossil cheetah specimens known is not great. At many localities only a single animal is represented, often by only one piece. The most significant exception is at the French site of Saint-Vallier on the east side of the Rhône Valley, where cranial remains of several individuals were recovered from deposits approximately 2.1 million years old. This general scarcity seems to fit what we know of modern cheetah behavior, however, since they are largely solitary animals except in the case of females with subadult young. The hunting method of a high-speed chase does not

FIGURE 3.34 *Life appearance of a leopard,* Panthera pardus

When we compare the leopard with the jaguar, seen in a similar pose in figure 3.47, we can appreciate the lighter build, longer legs, and relatively smaller head of the former, as well as the differences in coat pattern. The animal depicted here shows the pattern typical of leopards from Botswana, but some Asian individuals have larger rosettes that superficially resemble those of the jaguar.

Reconstructed height at shoulder 70 cm.

lend itself to cooperative action, and the number of cheetahs in a given area is therefore likely to be fairly small. This fact, together with the need for territorial separation, may in large part explain the wide geographic range of the cheetah, achieved at an early date and maintained over a considerable period of time.

MIRACINONYX. From North America come two fossil cats referred to the species *Miracinonyx inexpectatus* (figure 3.43 and plate 8) and *M. trumani* (figure 3.44). But the precise taxonomic status of these animals remains

unclear. The fossils range in age from perhaps 3.2 Ma for the larger *M. inexpectatus* down to the last 10,000–20,000 years for *M. trumani*. While both exhibit the general features and bodily proportions of the cheetahs, the former is broadly similar in size to the European Plio-Pleistocene *Acinonyx pardinensis*, while the latter bears greater resemblance to the living species. As a result, they have been placed by some authors in *Acinonyx*, by others in a subgenus of the cheetah, *Miracinonyx*, and by yet others in *Miracinonyx* as a distinct genus.

Although these animals exhibit the slim and elongated bones of the living cheetah, together with the shortened skull and narrow, high-crowned teeth, they differ in a number of detailed skeletal features. These features, which are said to include the retention of more fully retractable claws, have been taken to indicate that the American taxa exhibit a more primitive condition. That would imply that the American species have characters found in cats in general and would argue against an especially close link with the

FIGURE 3.35 *A leopard carrying the body of a hominid watched by* Deinotherium

This scene, set in the lightly wooded savannas of eastern Africa around 2.0 Ma ago, shows a leopard carrying the body of a robust australopithecine hominid, *Paranthropus boisei*—perhaps in search of a tree into which it can retreat from potential carcass thieves such as hyenas or lions. The extinct proboscidean *Deinotherium bozasi*, a distant relative of the true elephants, was a fairly common member of the large mammal fauna at the time, although it became extinct soon after. The leopard posed it no threat, but the giant animal is likely to have shown enough reaction to the presence of the cat to prevent the use of trees in the immediate vicinity.

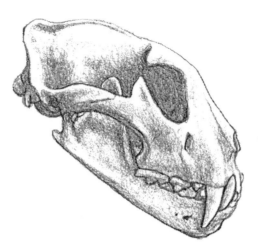

FIGURE 3.36 *Skull and life appearance of the snow leopard,* Panthera uncia
This is perhaps the most beautiful of all the living cats. However, the appeal of
its coat and the use of its body parts in medicines may yet cost the species its
very existence.

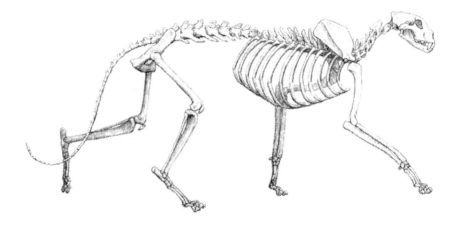

FIGURE 3.37 *Skeleton of the snow leopard*

The long, dense fur of the snow leopard gives the animal a rather stocky appearance, although it can be seen here that it has a rather gracile skeleton. The body proportions, with a small head and long limbs, to some extent resemble those of a cheetah; however, the metapodia, and especially the metacarpals, are shorter than in the cheetah. The excellent performance of the snow leopard in mountainous terrain may indicate that some extinct cheetah-like forms with similar proportions could also have done well in such regions. Some of the smaller cats, such as lynxes and servals, also exhibit very elongated limbs and small heads, although in each case these characters seem to have evolved independently.

Reconstructed height at shoulder 70 cm.

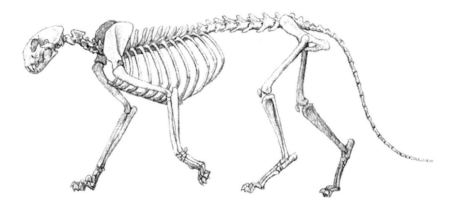

FIGURE 3.38 *Skeleton of a puma,* Felis concolor
Reconstructed height at shoulder 65 cm.

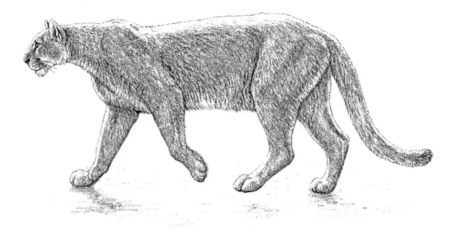

FIGURE 3.39 *Life reconstruction of* Felis concolor

Although the puma is often referred to as a "mountain lion," its bodily proportions are quite different from those of the true lion. The head is smaller and more rounded, the forelimbs are relatively shorter, and the long, densely furred tail lacks the black tuft at the tip.

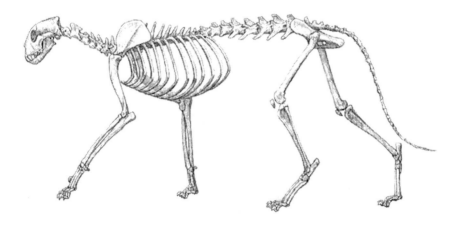

FIGURE 3.40 *Skeleton of* Acinonyx pardinensis

The skull of this large form of cheetah is well known from the French site of Saint-Vallier, but the best collection of postcranial bones comes from the slightly older site of Perrier in the Massif Central where long bones and much of the vertebral column of one individual have been found. This material shows that *Acinonyx pardinensis* was as specialized a sprinter as the living cheetah, with very elongated limbs. Unfortunately, the metacarpals were not recovered, and we have therefore reconstructed them with the same relative length as the living species. The back of the Perrier specimen is especially long.

The European cheetah has been said to be of lion size, and the Perrier specimen was indeed as tall at the shoulder as a small lion. But this is due to the elongation of the limb bones, and the living animal would have weighed much less than any lion.

Reconstructed height at shoulder 90 cm.

FIGURE 3.41 *Overall size comparison between* Acinonyx pardinensis (left) *and* Acinonyx jubatus

The specimen from Perrier is shown here in comparison with an average modern cheetah. Since the extinct form lived in generally colder environments than its living relative, it is reasonable to imagine that it developed a winter coat with rather dense fur, as leopards and tigers do today in China and Siberia. Such a coat would give the animal a stockier appearance, at least during the colder months. We have shown the Pleistocene species with a coat patterning similar to that of the so-called king cheetah, in reality a variant of the normal arrangement.

more derived morphology of the Eurasian cheetahs. The earlier of the American forms, *M. inexpectatus*, has also been seen in the past as morphologically close to the puma, indeed possibly even ancestral to it.

The question of claw retraction is dealt with in the next chapter, where this feature of the cats is examined in more detail. But even if the taxonomic distinction of the American and Old World cheetahs and cheetahlike cats is correct, there is no reason to imagine that they are unrelated. The appearance of the cheetah morphology in both Afro-Eurasian and American regions at about the same period, and the apparently less derived nature of the American forms, may argue for an origin in the latter continent followed by a dispersion across Asia.

FIGURE 3.42 Acinonyx pardinensis *in pursuit of* Gallogoral meneghini
Although it was considerably larger than the living cheetah, *Acinonyx pardinensis* may have found antelopes such as *Gallogoral meneghini* at the very top of its prey size range.

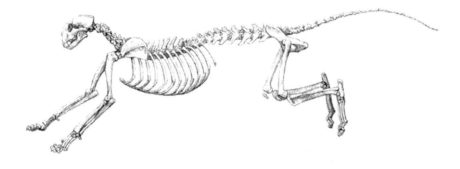

FIGURE 3.43 *Skeleton of* Miracinonyx inexpectatus

This restoration is based on an almost complete skeleton discovered in Hamilton Cave, Virginia. The body proportions of this early (Plio-Pleistocene) species of *Miracinonyx* are intermediate between those of a puma and those of a modern cheetah, and the animal is larger than either. The lower limbs are less elongated than in the cheetah, and the animal has fully retractable claws. The overall impression gained from this skeleton is that of a versatile animal, faster-running than the puma but stronger and better equipped for climbing than the cheetah.

Reconstructed height at shoulder 85 cm.

FIGURE 3.44 *Skull and head of* Miracinonyx trumani

The skull of *Miracinonyx trumani* resembles that of the living cheetah in being shortened and domed, with an enlarged nasal opening and rather short canine teeth. Like the cheetah, and unlike other cats, it also has reduced infraorbital foramina, which in the cheetah are associated with reduced whiskers.

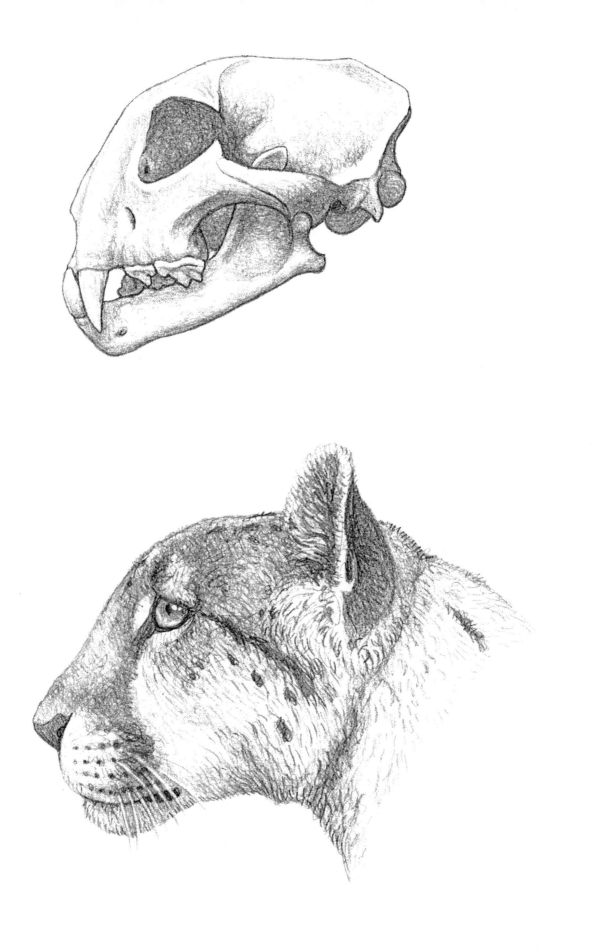

This is one of the latest species of *Hoplophoneus,* known from the Brule Formation of early Oligocene age. Like *H. mentalis,* it belongs to the group of "robust" *Hoplophoneus* species, and it attained a fairly large size. Its fossils are found in deposits that indicate rather open environments, but the bodily proportions of the animal are similar to those of living cats in forested habitats. We may imagine that, like living leopards, *H. occidentalis* would have been quite at home in vegetated patches and more forested ravines but perhaps capable of venturing into open spaces in search of prey. Other carnivorous animals from the same formation, like the hyaenodontid creodonts, were perhaps better suited to living permanently in open terrain.

We have shown *H. occidentalis* with a spotted coat, designed in accordance with the basic "aeluroid" pattern. Like true felids, these nimravids seem to have been dimorphic, and in the illustration we see the smaller female drinking from the pool.

PLATE 2 *Portrait of* Metailurus major

This cat was about the size of a large American mountain lion, or puma, and in fact may have resembled that animal to some extent. Postcranial bones from Pikermi in Greece and Montredon in France attributed to *Metailurus* are very feline in their morphology, as are some complete skulls from China and Greece despite some slight machairodont traits in their dentitions.

PLATE 3 *Portrait of* Dinofelis diastemata
This scene is set in wooded terrain at the
French early Pliocene locality of Perpignan.

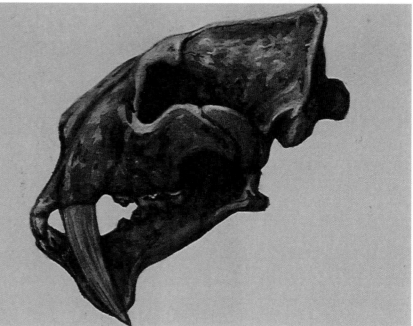

PLATE 4 Megantereon cultridens *skull and head* This reconstruction is based mostly on the skull of the complete skeleton from Senèze in France. *Megantereon cultridens* has the most developed mandibular flange among the felid machairodonts, and this gives the animal a superficial resemblance to some of the nimravids. African and European specimens of *Megantereon* exhibit the largest flanges, while in some Asiatic specimens they are small and rather similar to those of the early *Smilodon* species *S. gracilis*.

The skull of *M. cultridens* has some very derived features, such as an extremely well developed mastoid process, a lowered glenoid articulation, and a very reduced coronoid process of the mandible.

PLATE 5 Megantereon cultridens *with the carcass of a young fawn* Having caught its prey, the cat, a large male, turns and snarls at the suspected presence of a potential competitor.

PLATE 6 *Couple of* Smilodon populator
A couple of the South American
species are shown greeting each other
in open countryside. Note the massive
forequarters of the animal.

PLATE 7 *Scene with lions and mammoths in Alaska during the upper Pleistocene*
Prey species in Africa today appear to judge whether a group of lions is
interested in hunting or merely moving to another part of the pride territory, and then to act accordingly. While adult mammoths would have
been immune to predation, the lion may have posed a sufficient threat
to young or sickly members of the herd to cause agitation among the
adults and to have ensured that a constant watch was kept on any movements involving more than a single pride member.

PLATE 8 *Portrait of*
Miracinonyx inexpectatus
From the range of different
habitats occupied by this
species in the Plio-Pleistocene
of North America we have
chosen to depict it in the hilly
landscape of what today are
the Appalachian mountains.
It also lived in coastal savanna
environments in Florida and
in open forests in Arkansas.
The versatile anatomy of this
cat, less specialized than the
later species *Miracinonyx tru-*
mani or the Old World cheetahs, undoubtedly helped it
to survive in such a range of
conditions.

PLATE 9 *Two inexperienced*
Homotherium *individuals*
attempting to kill a horse

This picture shows the kind
of attack that a lion or a tiger
might be able to attempt but
that would be extremely dangerous for any cat with long
upper canines. Unless a rapid
severing of the major blood
vessels were achieved, any violent movements by the horse
would almost certainly lead to
damage to the teeth.

PLATE 10 Homotherium serum
with a Dall sheep ram

A wide selection of prey species
was available to *Homotherium
serum* in the mammoth-steppe
environment of Alaska during
the upper Pleistocene. It may
therefore have been advantageous for this cat to have a
whitish coat, thus rendering it
less conspicuous in snowy conditions. In wild populations of
lions and tigers today white
mutants (as distinct from albi-
nos) occur from time to time,
and although they rarely survive to breed, selection pressure during the past may have
favored such variants.

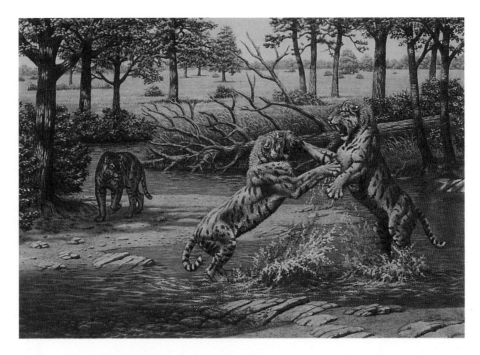

PLATE 11 Machairodus giganteus *males fighting*

Machairodus giganteus appears to have been a highly dimorphic species, and some of the skulls and limb bones found in China and Europe show that males often attained very great size. It is reasonable to imagine such males fighting for territory and thus for access to females, as depicted in the illustration.

M. giganteus is associated with faunas typical of open plains in the Turolian of Eurasia, and its lifestyle may have resembled that of the lion to some degree.

PLATE 12 *Group of* Smilodon fatalis *hunting a bison.*

Even with the strength of *Smilodon*, the capture of an adult buffalo probably required the collaboration of several individuals. The species depicted, *Bison antiquus*, is the most commonly found herbivore in the Rancho La Brea deposits and was very likely a common prey of *Smilodon*.

PLATE 13 Machairodus aphanistus
pursuing Miotragocerus

A very abundant species in Vallesian
(upper Miocene) sites of Europe,
the antelope *Miotragocerus pannon-
iae* was probably a favorite prey for
the tigerlike *Machairodus*. The hoof
structure of the antelope suggests
that it was probably slow-moving
compared with the three-toed horse
(*Hipparion primigenium*), but that it
may also have been a fairly profi-
cient swimmer. This implies that,
like living aquatic antelopes of the
genus *Kobus* (water buck),
Miotragocerus probably took to the
water for safety. *Machairodus
aphanistus*, in turn, would have tried
to catch individuals unaware and
cut off their retreat to safety.

PLATE 14 *A Mid-Miocene landscape in the Vallès Basin of Catalonia*

The scene depicts a riverine forest with *Chalicotherium* and the proboscidean *Deinotherium,* the hornless rhinoceros *Aceratherium,* the suid *Listriodon,* and the small cervid *Stephanocemas.*

PLATE 15 Homotherium serum *in the snow*
Although the North American representatives
of *Homotherium* seem to have been less abundant than *Smilodon*, they were successful as
judged on the basis of their longevity: they
went extinct only at the end of the Pleistocene.
They seem to have occupied a range of higher
altitudes and latitudes than *Smilodon*, and were
probably well adapted to cold environments; it
would be reasonable to envisage them developing long winter coats as lynxes and snow
leopards do.

Here a mother is shown leading her adult-
sized cubs, but it is not impossible that the
species had more complex social groupings.

PLATE 16 *Landscape at Pontón de la Oliva*
Giant hyenas, *Pachycrocuta brevirostris*, are
shown resting during the heat of the day,
only to be disturbed by three dhols (*Cuon
alpinus*) running past their shelter.

CHAPTER 4
Anatomy and Action

I N THIS CHAPTER WE TRY TO BRING TOGETHER WHAT WE KNOW OF some of the most important and distinctive features of the anatomy of the living cats, and to show how their mobility and range of activity is intimately connected to their physical attributes. Based on this understanding of anatomy and action in the living species, we may then begin to make inferences about the capabilities of some of the fossil species.

We begin by examining the question of coat color and patterning, since this is one of the most obvious features of the appearance of the cats.

COAT COLOR AND PATTERNING

Coat color and patterning among the living cats appear quite variable, but there are some general rules and a few broad categories into which we can place most of the animals. Essentially, most cats show a fairly typical pattern of spots, rosettes, or stripes on a yellow background of varying intensity. As we see in figure 4.1, members of some species, such as the lion or the American puma, have a largely uniform coloring when adult, although the coats of the young are patterned and may still display such markings well into the early stages of adulthood. In contrast, very young cheetahs show relatively little patterning while the adults are strongly marked. One strange feature, often seen in leopards in particular, is the melanistic or black variant (frequently but wrongly considered to be a separate species, the black panther). This variation also appears with some frequency in the jaguar population. Even in these melanistic animals, however, the underlying pattern can usually be seen if the animal is viewed from a suitable angle so that the light picks up the arrangement.

The genetic basis of coloration seems to be generally well understood. It has been suggested that all the coat markings may have derived from a basic pattern of darker stripes that broke up to form smaller spots and

FIGURE 4.1 *Adults and juveniles of cheetah* (top), *puma* (center), *and lion*
This illustration shows the differences between the juvenile and adult coats of
three living species.

rosettes in some and simply faded in others. These darker areas show up on a yellowish background of varying intensity, and if the genetic coding for the yellow background becomes altered then the result is a largely or entirely black individual. However, it is worth bearing in mind the fact that the young of some species in which the adults are unmarked, such as the lion and the puma, go through a spotted phase. This may point to the fact that the primitive condition was in fact spotted.

It seems likely that coat patterning has evolved at least in part under selective pressure for concealment, and it is certainly true that leopards and jaguars, for example, can be extraordinarily difficult to see in the dappled shading of a tree even when their presence is known. But the probable importance of the patterning to the animal itself, as a means of mate recognition or more general social signaling, should not be overlooked. The points we made earlier about the fertilization system providing the coherence of the species might be significant here, and the detailed patterning of some felid coats could be extremely relevant. The distinctive black mane of male lions seems an obvious example of a pelage (hairy covering) feature with clear social significance as an indicator of status within the pride (figure 4.2). Other details, such as the white spots behind the ears of tigers and the white tip to the tail (or underside of the tip) in leopards and cheetahs, have been plausibly suggested as important visual clues to young animals following their mothers through long grass.

In our reconstructions of the living appearance of animals known only from the fossil record, the question of coat patterning is clearly an open one. In each case, what we have tried to do is to depict the animal with a suitable pelage, rather than leaving them all a neutral color that would itself be equally open to question. We have also taken account of the wide distribution of marking, and the indications that this was the primitive, and therefore widespread, condition (see the section on reconstruction, below).

Of course the coat is in fact a pelage, a fur made up of hairs. It is the individual hairs that are colored and produce the patterning, but in addition to color variation the hairs of the coat also differ in length and thickness according to position. The most noticeable difference is in the whiskers, highly modified hairs, whose function is considered in the next section.

SENSES

Good sensory abilities are essential for many animals in a variety of circumstances, but particularly so for predators who have to locate and bring down prey that may be actively seeking to avoid capture and death. In the cats some sensory abilities appear to be more highly developed than others, although this is a difficult area in which to talk about absolute differences

FIGURE 4.2 *Adult and subadult male lions*
Adult male lions, like the one shown at left, reaffirm their dominant status with a display of body size and mane. The display incorporates a stiff standing or walking posture called "strut," which exaggerates the differences between the adults and subordinate juveniles such as the animal at right.

and to point to certain animals as having the "best" eyesight, the "best" hearing, and so forth.

Eyesight

Domestic cats are famous for their ability to see in the dark. This fame is based on fact, and stems from important differences in construction between their eyes and those of humans. However, not all cats are equally well equipped for night vision, while other species besides cats may also see well in the dark.

At its simplest, enhanced night vision depends upon being able to get more light into the eye so that it can be detected by the optical nerves that relay the information to the brain (this principle underpins all optical devices designed to improve our own ability to see in the dark). At the same time, unless the animal is to be entirely nocturnal, the eye must retain the ability to operate during the day when vastly greater levels of light are available. These competing requirements are met by a series of modifications that permit various amounts of light to be admitted to the eye and that enhance the very lowest levels (figure 4.3).

The first element, variation in the amount of light admitted, is achieved by having a pupil that opens and closes. The pupil is the dark center of the eyeball, and its ability to open and close may be easily seen by standing close to a mirror in strong light. Close both eyes for perhaps ten seconds, and then open them and immediately look closely at the pupils in the mirror.

FIGURE 4.3 *The mechanism of pupil closure in the cat*
The figure shows how the pupil of many cat species can be closed to a slit when pulled by the ciliary muscles. In darkness (*left side of left drawing*) the pupil of a small cat of the genus *Felis* is almost circular in shape. In bright light, the pupil contracts to a mere slit (*right side of same drawing*). The muscles that control the pupil (*shown in the righthand illustration*) are drawn across each other, instead of being arranged in a circle around the pupil as in the case of our own eyes.

You will see them reducing in size, having opened while the eye was closed in an effort to admit more light.

While we have round pupils, the domestic cat has a slit pupil. The advantage of a slit is that it can close the aperture down beyond the point achieved by a round one, and if necessary it can close completely, thus providing maximum opening and closure to take account of all conditions. In this way the eye is protected against too much light. Many of the smaller cats have such a slit pupil, but it is less common in the larger species, and in the lion it is only slightly oval in shape.

The extent of night vision is determined by the sensitivity of the retina, a layer of light-receptor cells on the back surface of the eyeball. This layer is connected, via the optic nerve, to the brain, and it is here that whatever light has been admitted is detected. To cope with low levels of light, the eyes of nocturnal animals have yet another refinement in the form of a reflective layer behind the retina, known as the tapetum lucidum. This layer returns light that has passed through the retina, giving a double opportunity for detection. It is this reflection that produces the well-known effect of eye-shine when a light is thrown on nocturnal animals in darkness: what is seen is the light being returned from the tapetum layer.

The fact that only the smaller cats have a slit pupil, taken by itself, might suggest that it is only these species that operate at night with any consistency, and that their night vision is better than that of their larger relatives. However, it is clear that many of the larger species do operate outside daylight hours with some regularity, and that the extent of night vision is evidently adequate for this purpose.

In addition to being able to see in various conditions, the eyes have to provide information on distance. This is especially important for animals that live by catching fast-moving prey, often by leaping on it. The best way to obtain distance information is by a system of binocular vision, in which the fields of view from the two separate eyes overlap to some extent and allow the brain to take account of both pieces of information in judging how far an object is from the observer. Such a system in turn requires that both eyes face forward, as do our own, and not to either side. Eyes that can face in different directions may, of course, have other advantages, such as giving a wider overall field of view, but this can also be achieved in forward-facing eyes by increasing eye mobility. Among the Carnivora, the cats have the greatest degree of binocular vision, as might be expected.

Of course we have little direct evidence of eyesight in fossil cats. But we can observe the size and arrangements of the orbits (eye sockets), and deduce from this that binocular vision was present. Many of the machairodont cats appear to have had rather small eye sockets, implying that most of their activities were concentrated into daylight hours. The exception to this may have been later members of the genus *Homotherium*, in which the orbits seem to have been somewhat enlarged. We also have some evidence bearing on the question of vision from the size and organization of the brain (figure 4.4), since the inside of the skull gives us some indication of the brain's superficial arrangements. Later *Homotherium* in America appears to have had an unusual enlargement of the visual cortex, that portion of the brain which in modern cats deals with processing visual information.

In contrast to the machairodonts, earlier ancestors of the living large cats appear to have been relatively large-eyed animals, implying an important role for nighttime activities.

Hearing

Relatively few studies appear to have been undertaken on the hearing abilities of large cats, although the naturalist George Schaller did suggest that lions seem perfectly able to hear sounds over considerable distances. Some work has been reported for the domestic pet and suggests an ability to hear

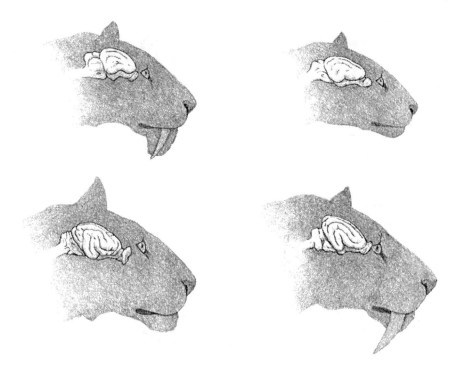

....

FIGURE 4.4 *Profiles of cat skulls with brain shapes*

The shapes of the brains of many fossil cats are known thanks to natural endocasts (where cemented sediments in the brain cavity preserve the details of the inner surface of the skull) or to casts taken from well-preserved skulls. In this figure we see the brains of a nimravid, *Hoplophoneus* (*above left*), and three felids in the form of *Proailurus* (*above right*), *Panthera* (*bottom left*), and *Smilodon* (*bottom right*).

Both the nimravid and the very early felid *Proailurus* show a simpler pattern of sulci (furrows) and gyri (convolutions), implying less brain complexity. *Smilodon* displays a modern sulcal pattern, just like *Panthera* and all other felids of post-Miocene age. Compared with the primitive condition, the brain of modern felids has increased sulcal complexity in the regions that control hearing, eyesight, and limb coordination.

considerably beyond the upper limit for humans, although this enhanced ability declines with age. The external ear of the cats, known as the pinna, is composed of cartilage under the skin and is quite mobile in a way that gives the impression of a scanning device attempting to track down the position of sounds. The relative positioning of the pinnae on the head of various species may have some functional relationship to the method of prey capture, permitting finer discrimination of sound levels and origin (figure 4.5).

Because cartilage does not fossilize we have no direct information on the shape of the external ear in extinct cats, but the structures of the middle

FIGURE 4.5 *The mechanism of the ear and the positioning of the external ear*
In the first drawing we see the head of a leopard with the structure of the external ear and the main muscles that move it. We have shown the cartilage of the ear (or auricle) without hair, to illustrate the bursa and the incisura intertragica (lower-most part of the external ear opening). As can be seen, the muscles controlling the auricle come from all sides, permitting a wide range of movement. In the second drawing we show the ear covered with hair and in a different position, this time completely turned back.

The third and fourth drawings are frontal views of two modern cats, the serval on the left and the ocelot on the right, to illustrate the possible range of variation in positioning for the auricle. The two animals are of similar size and the rear parts of their skulls are not markedly different, but the ears of the serval nearly touch each other on the top of the head while those of the ocelot appear much more laterally placed. In fact, the main difference is in the size of the auricle, and the incisura intertragica would seem to be in a more or less fixed position relative to the ear opening of the skull.

and inner ear within the skull itself, and the positioning of the opening, suggest no major differences from living species in terms of hearing abilities.

Smell

Cats, like most carnivores, are equipped with odor-producing glands, and the odors appear to form a significant component of social signaling and discrimination. The marking of territory—by spraying urine containing anal gland secretions in the case of males, or by rubbing objects in order to deposit odor from glands of the head—is a feature of behavior among all members of the family, and implies a considerable level of discriminatory ability (figure 4.6). However, cats generally have reduced nasal areas in comparison with other members of the order such as dogs, and Leonard Radinsky has reported finding relatively smaller olfactory regions in the brains of cats (relative, that is, to body size) than in both canids and viverrids.

One particular aspect of smell that does seem to be especially important for cats involves the characteristic grimace known as "flehmen," in which the nose is wrinkled and the lips pulled up. This is commonly done by males during courtship, and appears to be an attempt to gauge the likely receptivity of the female in estrus by detecting chemical signals in the urine. Jacobson's organ, a pouchlike structure at the front of the roof of the

FIGURE 4.6 *The sense of smell: A snow leopard sniffing the marking scent of another animal*
The cat is shown adopting a position typical for most cats when they encounter marking scents. It will probably follow this by marking the spot with its own odor: by rubbing its head and neck on the branch, by spraying urine, or by doing both.

mouth, lined with receptor cells like the nasal apparatus and connected by a duct to the mouth itself, has been implicated in this behavior, the facial expression perhaps serving to bring the odor into more direct contact with the organ.

Touch

It may seem that to the cats, with eyes and ears capable of seeing and hearing in most conditions, allied to a good sense of smell, touch would be of relatively little importance. But like many carnivores they have well-developed whiskers, or vibrissae as they are correctly known, and the size of these suggests that they are of some importance. This seems to be the case for mammals in general; for example, rats placed in a maze, even in good light, are apparently lost if their vibrissae are trimmed. We mentioned above that the vibrissae are actually modified hairs, and although they may be present on other regions of the body they are most commonly considered as cranial features. In cats they are present above the eyes and on the sides of the head, but to many observers the whiskers of the upper lip, known as the mystacials, are the most obvious.

The whiskers are extremely sensitive to touch, as may be readily seen in the domestic cat. The individual hairs are very enlarged and stiff, and their bases in what is known as the vibrissal pad are richly supplied with sensory nerves. Paul Leyhausen, in his classic study of cat behavior, mentions how blindfolded domestic cats were still observed to seize and kill mice with a precisely directed bite once their whiskers made contact with the prey; further, they were unable to direct their biting without the whiskers while blindfolded, even though their sighted ability to do so was unimpaired. Leyhausen also provides pictures of cats attacking a bird and carrying a mouse, and in each case the whiskers are clearly employed to sense the precise positioning of the prey (almost enveloping the prey in the case of the carried rodent). It seems clear that, for the domestic cat at least, the whiskers are an important addition to its sensory equipment. But what of, say, a lion, a leopard, or a tiger? It seems hard to imagine that the possession of whiskers has much to offer in the capture of a buffalo, a gazelle, or a deer, and Leyhausen's conclusions in this area are not entirely clear. However, the size of the whiskers in the larger cats, and the extent of innervation to them, suggests that they too can convey considerable sensory information (figure 4.7). Precision in biting during the final stages of the kill may of course be an important factor even for large cats (as discussed in more detail later in this chapter and in chapter 5), especially at night.

Unfortunately, whiskers do not fossilize. This is a great pity, because they lend much of the character to the face of a cat and are therefore important components of any life reconstruction. But we do have a couple of clues. First, the incidence of whiskers is high among the Carnivora and among living Felidae, indicating that it is probably a primitive condition shared with extinct members of the order or family. Second, if it makes sense for living cats to make precise bites, then it would be at least as logical for the extinct machairodont species to have had a well-developed sense of where the prey was in relation to their teeth, in order to avoid damage when biting. The observation that domestic cats are able to orient themselves correctly for a neck bite on a mouse while blindfolded is therefore of considerable general significance. And third, and more directly, the sensory nerves that serve the cat's whiskers pass through the infraorbital foramen, and represent the majority of the nerves that do so. We can therefore infer that any cat with a well-developed infraorbital canal probably had a well-developed cluster of nerves passing through it to the base of its whiskers. Such a canal is certainly present in the fossil skulls that are available, and especially in some of the machairodont taxa such as *Smilodon* and *Megantereon*. We have therefore shown fossil and living cats with equally well developed whiskers.

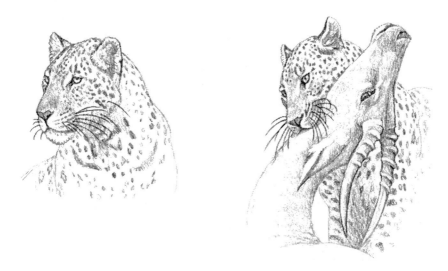

FIGURE 4.7 *The arrangement and use of the whiskers*
The drawing on the left shows a leopard in a relaxed pose. The whiskers are extended laterally or even slightly backward. The animal on the right extends its whiskers in front of its mouth as it applies a throat bite to an impala, and by doing so it is able to make a more precise judgment of the movement and position of the prey.

BONES AND BODY PROPORTIONS

All cat skeletons, apart from their size variation, are broadly similar, and we have already pointed out that a lion is in many respects essentially a scaled-up version of a domestic cat. If you had a labeled skeleton of a domestic cat, you could easily use it to sort a jumble of lion or leopard bones into the same skeletal parts, although you might notice some differences in the proportions of the various limbs and their constituent parts. Those differences reflect differences in lifestyle and mobility, and are important factors in any effort to reconstruct the behavior of extinct cats.

Most cats, living and extinct, have elongated and relatively slender long bones (as the main bones of the limbs are often known) and elongated foot bones in comparison with some other animals of similar body size. Cats are said to be cursorial (running) carnivores, like dogs and hyenas (figure 4.8), as opposed to ambulatory (walking) species like the bears, as illustrated in figure 4.9. The distinction refers to the habitual gait of the animals, and of course cats may walk at times, while bears will certainly run (indeed, many people have ended their lives being surprised by just how fast bears can run). The term "cursorial" thus covers a multitude of different locomotory abilities, but it serves as a broad distinction. Most of the cats are capable of achieving high speeds at least for short distances, often with considerable acceleration from a standing start, and the skeletal proportions generally betray this fact. The ability of the cheetah to chase at high speed for some several hundreds of meters is extreme. The foot bones of the cats are closely bound together by ligaments, and are thus able to withstand the forces generated during the landing phases when running at high speeds.

The tail is frequently long, and for the cheetah it is an important means of maintaining balance during the high-speed twists and turns of a typical chase (figure 4.10). The high running speed of the cheetah is further aided by its flexibility: it is built more like a racing greyhound than any of the other cats. Its lean body and long limbs permit it to increase its stride length by flexing its back, so that in its fastest running phase, traveling at perhaps 90 km/h (precise figures conflict; this may be a minimum estimate), it covers about ten meters with a single leap. At that speed it is not so much running as proceeding by a series of long and rapidly repeated bounds in which the body coils and stretches. (See the reconstructed running sequence of *Miracinonyx* at the end of this chapter.)

The variations in body shape that we see between the various cat species are also reflected in the skeleton. The limbs of the fast-running cheetah are much longer and more lightly built than those of a leopard or a jaguar, especially in the forearm and shank, and less massive than those of a tiger or a lion. A typical cheetah may have a radius/humerus length ratio of 1.0, versus around 0.9 in a jaguar or leopard, giving it a longer "forearm" in relation to its size (figure 4.11). All the cats also increase their effective limb length by what is known as a digitigrade stance—that is, they stand on their

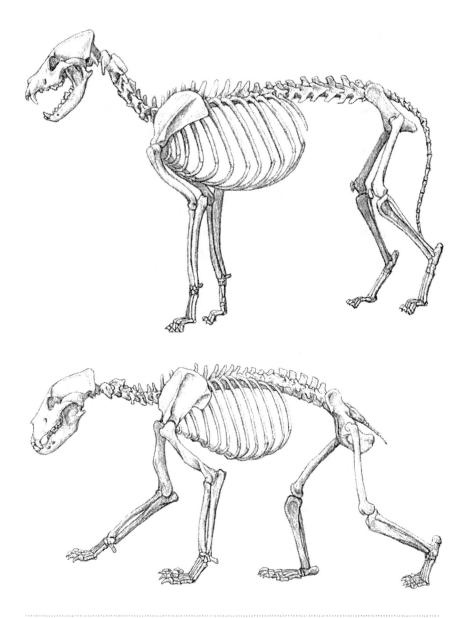

FIGURES 4.8 AND 4.9 *Skeletons of a wolf,* Canis lupus (top), *and a brown bear,* Ursus arctos
These skeletons illustrate the differences between the body shape and proportions of the cats and another cursorial hunter, the wolf, in comparison with a bear. The wolf has bones that are recognizably similar to those of a cat of equivalent size, but the back is shorter and the head is proportionally longer as a result of a longer tooth row. The scapula of the wolf tends to be elongated and its movement thus adds to the stride length (a feature repeated in the scapula of the cheetah), while the longbones tend to be rather more slender. The stance of the wolf, like that of the cat, is fully digitigrade, although a more detailed study would show that the paws of the cat are capable of a greater range of movement.

In contrast to both, the bear is stocky and robust, with enormous muscle insertions on the scapula and humerus in particular. The skull and jaws are massive, and the cheek teeth are broad and adapted to coping with a range of food items. The back is very short and the stance is plantigrade, although the forefeet do not rest completely flat on the ground.

toes, and the bones in their feet equivalent to those that form our palms and soles are greatly elongated. This feature they share with the cursorial dogs and hyenas, in marked contrast to the ambulatory bears and of course ourselves, with our plantigrade stances that involve placing the whole of the foot (especially the rear foot, in the case of the bears) on the ground with every stride. Not surprisingly bears, like us, have short bones in the palms and soles of their fore and hind feet. Some of the differences in these stances are highlighted in figure 4.12.

When we look at the fossil cats, we see a number of differences from the modern condition. Although we have no direct evidence for much of the body plan of *Proailurus*, the fact that the later *Pseudaelurus* retains the long, flexible back of more primitive carnivores implies that *Proailurus* was little different. The earliest cat was therefore probably a very able climber, like the living arboreal viverrids such as the genets, the palm civets, and the fossa of Madagascar (figure 4.13), thanks to its small body size and viverrid-like proportions. The long hind limbs would have provided efficient propulsion, while the short radius and metapodia would have given the animal grasping strength. A plantigrade stance, with a large surface of contact between foot and branch, is safer in the trees than a digitigrade one, and a wide range of movement at the ankle and the wrist joint allows the position of the feet to adjust to irregular surfaces.

All the earlier carnivores share these adaptations to a life spent largely in the trees. As the cats evolved, they departed from this primitive condition to varying extents as they developed more efficient forms of terrestrial locomotion. Thus the metapodia of *Pseudaelurus* show a move toward elongation, while the ankle joint permits less lateral rotation than was possible for *Proailurus*. This latter trait is important, because as the lateral rotation of the joint is reduced, movements in the vertical plane become more efficient. This is the kind of movement best suited to locomotion on land.

Among the "normal" cats there is an overall pattern of body build that is similar to that of *Pseudaelurus*, but with a series of differences pertaining to increases in size and the growing importance of terrestrial locomotion. Large size in cats necessitates a stronger back, if only to support the weight of the viscera, and the back must therefore be relatively shorter. Hence most of the cats larger than a lynx, and especially the members of the genus *Panthera*, have shorter backs than *Pseudaelurus*. Some of the extinct cheetahs seem to have the longest backs among the large felines, as long relative to their size as in *Pseudaelurus*.

Smaller cats retain long backs, but usually differ from the pseudaelurine condition in having a longer radius and metacarpus, both of which are adaptations to terrestrial locomotion. We see this tendency even in the partly arboreal ocelot. In contrast, some of the larger cats that live in dense forest, such as the clouded leopard and the jaguar, have quite a short radius and metacarpus. This may at first glance appear to be a primitive retention

FIGURE 4.10 *A sketch of the use of the tail by the cheetah*

This cheetah, turning at top speed, uses the motion of its tail to counterbalance its body weight. Similar use of the tail may be seen in the domestic cat when jumping or walking on a narrow object such as a wall.

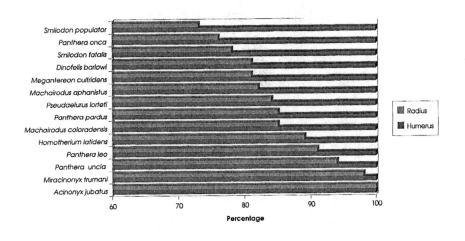

FIGURE 4.11 *An illustration of some forelimb segment ratios*

The diagram shows the relative proportions of the humerus and radius in a series of living and extinct felids. Each species is shown with the humerus scaled to the same length, and the radius length expressed as a percentage of the humerus measurement. Note how in the cheetah (*Acinonyx*) the two bones are of equal length. Typical specimens of each species are shown, but some individual variation within each taxon should be expected.

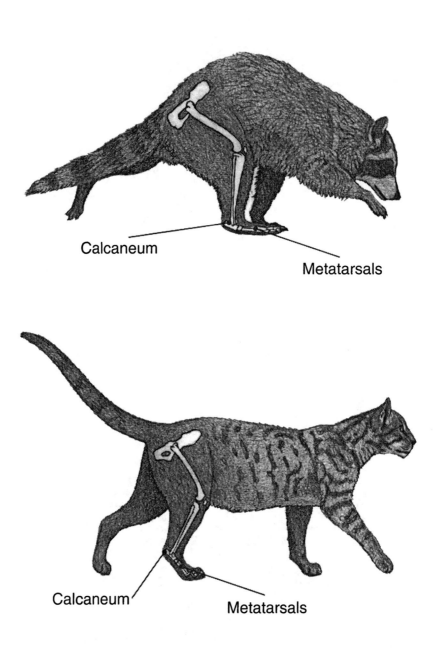

Calcaneum

Metatarsals

Calcaneum

Metatarsals

FIGURE 4.12 *Digitigrade and plantigrade stances*

The drawings show a plantigrade carnivore, the raccoon (*top*), and a digitigrade carnivore, the domestic cat, in a walking gait. The bones of the hind limb are highlighted in both animals to show the differences in stance. When stepping, the raccoon rests the metapodia flat on the ground, which is a plantigrade stance similar to our own, but unlike humans it always keeps the calcaneus, or heel, at some angle to the ground. In contrast, the cat rests only its "fingers" on the ground, with the metapodia lifted and the calcaneus completely clear; fibrous pads cushion the impact of the fingers and sesamoid bones (the small bones that take the tendons across the joints between metapodia and phalanges) on the ground.

FIGURE 4.13 Proailurus *in a tree in a fossalike pose*
Note the gain in stability from placing the rear foot flat on the branch.

from the ancestral condition, but it is more probably a secondary development from ancestors with longer limbs. In the particular case of the jaguar
the fossil record seems to confirm this latter interpretation, since Pleistocene specimens had longer metapodia than their living relatives. For the
living jaguar, and for the clouded leopard, it seems that strength is more
important, and speed less so, than for other large cats. On the other hand,
cheetahs are built for speed over short distances, but their body proportions can be derived from those of the puma with only slight changes. The
most important of these involve the elongation of the lower limbs—mostly
the radius and the metacarpus, although the whole limb is actually lengthened; for instance, in many cat taxa the length of the humerus is about ten
times the average length of the thoracic vertebrae 3 to 8, but in the cheetah the humerus is more than thirteen times that figure. The long back of
the cheetah is itself a retention for speed. Other carnivores that run to
catch their prey, such as wolves and hyenas, often do so over long distances
at a slower average speed, and have short, relatively stiff backs, which are
less energetically costly for such activity.

The machairodont species have their own pattern of differences from
modern cats, although the earliest members of that line share more common felid features. *Pseudaelurus quadridentatis,* for example, the likely
ancestor of the subfamily, had a very feline body plan, with rather primitive

traits such as a long back, short metapodia, and hind limbs longer than the forelimbs. By the later Miocene, we find animals like the small *Paramachairodus*, which had hardly changed that basic design, alongside animals like *Machairodus aphanistus*, which was the size of a lion and had elongated metapodia (figure 4.14). The size and proportions of the latter may point to a move toward a more terrestrial lifestyle.

By the later Pliocene, around 3.0 Ma ago, this pattern of a smaller and a larger species appears to have been well established, with the presence in many regions of the world of the lion-sized and long-limbed *Homotherium* together with the smaller and somewhat jaguarlike *Megantereon* (figure 4.15). However, these later cats were by then very different from the Miocene taxa. Both had very much shortened lumbar regions in their spinal columns, elongated necks, and shortened tails; but while *Megantereon* had become very robust in its limb proportions, with short legs of roughly equal length, the long hind limbs of *Homotherium* were matched by even longer forelimbs.

These body proportions of the later machairodonts require some explanation. As we shall see in the next section, the peculiar problems posed by the development of long upper canines would have meant that great care had to be exercised in dealing with prey carcasses. A long and powerful neck would have made it easier to reach vital parts once the prey was securely held, and we see parallels to this shared in saber-toothed taxa as diverse as *Thylacosmilus*, *Hoplophoneus*, and *Smilodon*. In contrast, the shortened lumbar spine appears at first sight to confer fewer advantages, reduc-

FIGURE 4.14 Paramachairodus (*left*) *and* Machairodus *together to scale*
Note the vastly greater size of *Machairodus*.

FIGURE 4.15 Megantereon (*left*) *and* Homotherium *together to scale*
The size difference here is much less extreme than in the case of the Miocene
pairing depicted in figure 4.14.

ing mobility and top speed and certainly making acceleration from a standing start rather more difficult. However, a shortened back would certainly
have been stronger, and the overall impression gained from the skeleton of
Megantereon is one of great power, giving little chance of escape to any prey
once captured.

The elongated neck and shortened lumbar region of *Homotherium*,
together with the elongated forelimbs and a high scapula, give the animal
a certain similarity to the shape of a hyena (figure 4.16). This apparently
ungainly stance was formerly exacerbated in reconstructions by the recognition of a somewhat shortened calcaneum (the ankle bone to which the
Achilles tendon that extends the foot attaches, forming what we call our
heel), a feature that may have reduced the cat's ability to leap to the extent
possible in modern cats. This shortening, together with other features of
the hind limb, was once thought to indicate a plantigrade stance for the
rear foot in this animal, which would certainly have given it a very peculiar
appearance. But even without such exaggeration, the stance is odd.

An interesting parallel is also seen among late Pleistocene *Smilodon populator* from eastern South America, which had a longer humerus and femur
in conjunction with shorter metapodia than its northern and western relative *S. fatalis*, and also a relatively longer front limb (figure 4.17). Some
authors have suggested that such proportions, seen in some larger African
antelopes such as wildebeest, for instance, make for easier cantering. It is certainly true that spotted hyenas, for all that their gait looks awkward, are able
to sustain a chase for long distances. This interpretation, if correct, might

FIGURE 4.16 Homotherium (*top*) *and* Crocuta crocuta *to scale*
The typically slope-backed appearance of the spotted hyena is seen to be paralleled in the cat.

imply that both *Homotherium* and *Smilodon* operated as rather long-distance hunters, at least for part of the time. In contrast, the American paleontologist Viola Rawn-Schatzinger has pointed out that the American *Homotherium* from Friesenhahn Cave in Texas has these peculiarities in conjunction with reduced claw retraction, features that in her view suggest a sprinting ability. We shall discuss some of these conflicting ideas more fully in chapter 5, when we consider the likely hunting behavior of some of the extinct cats.

The skull of the cats differs from that of many other carnivores in having a shortened face, although there are considerable differences between species. Leopards and jaguars of similar sizes have quite distinguishable skulls, because that of the jaguar is broader with orbits set farther forward. In the case of the cheetah, the skull is even shorter and more highly domed, with an enlarged nasal opening that permits the rapid inhalation of a great volume of air during and after the exertion of a chase. In comparing skulls from smaller to larger species it is also possible to observe differences that are in the main size-related. For example, smaller species have relatively larger eyes because there is a broadly optimum size range for this organ; larger skulls therefore have relatively smaller eye sockets. At the same time, the brain does not usually increase in size at the same rate as the body between smaller and larger species, so that the braincase of the skull becomes relatively smaller; however, the muscles that operate the jaw have to attach to the skull, and a smaller braincase means that extra crests

FIGURE 4.17 Smilodon fatalis (*top*) *and* Smilodon populator *to scale*

As can be seen, the back of the larger, South American species is distinctly more sloping and hyena-like.

of bone have to develop to provide the anchorage. The outline of the skull therefore changes across the size range for purely mechanical reasons (figure 4.18).

The greatest differences in skull morphology are those between the true cats of the subfamily Felinae and the saber-toothed species of the sub-family Machairodontinae. Many of those differences are functionally related to the possession of elongated canines, and are paralleled in the morphology of skulls in the Nimravidae and in other saber-toothed species among the marsupial carnivores. Before discussing these differences it will therefore be useful to consider the question of dental morphology in the Felidae.

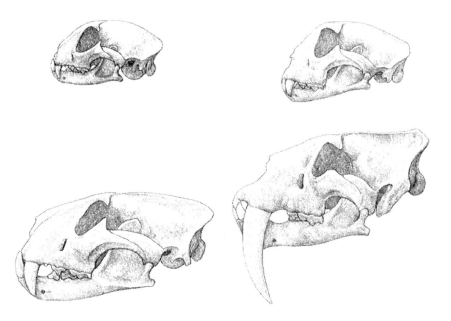

FIGURE 4.18 *Comparison of skull shape in some living and extinct cats*
In small cats, such as the serval (*above left*), the orbits and braincase are relatively very large. The muzzle is short and the sagittal crest very small or absent.

The cheetah (*above right*), in spite of being a large cat, has a relatively small head with a short muzzle and a poorly developed sagittal crest, although the fore-head is somewhat inflated by the enlargement of the sinuses.

In the largest individuals of the leopard (*below left*) the sagittal crest is so developed that the dorsal crest looks almost straight. The large canine roots contribute to the size of the muzzle.

In advanced saber-tooths such as *Smilodon* (*below right*), the features related to large size are combined with others related to the peculiarities of their biting method, further explained in figure 4.27.

TEETH

A great deal has been written in the scientific literature about the teeth of the cats, mainly provoked by efforts to understand the evolution of the extreme specializations seen among the saber-toothed species.

Since the cats are members of the Carnivora, their teeth exhibit many of the general features of the order, as may be seen in figure 4.19. But in comparison with, say, the dogs, all the cats have a very reduced dentition—that is, they have fewer teeth. That reduction comes about through specialization: whereas a dog is able to slice through meat and other relatively soft tissues and also retains the ability to crack bones, the cat is really only able to slice its food. Watching a domestic cat can be instructive here. Notice how it differs from a dog in the way it deals with its food, taking items from the bowl and chewing through pieces of meat on one side of its mouth, whereas the dog tends to put its muzzle in and bolt the lot down. You cannot feed a cat from the hand in the same way that you can a dog because each deals with its food in a different manner, and the cat usually drops the item to the floor before tackling it.

A dog will of course spend more time on eating if the food is hard or tough, or if it comes in larger pieces. Bones will be cracked and parts eaten, depending on the circumstances and the extent of hunger; wolves in the Canadian Arctic may return to a long-abandoned carcass in times of food shortage. Hyenas, especially the large spotted hyena, are even more adept at consuming bone and even better equipped to deal with it: their teeth permit them to break all but the largest items, and their digestive system can extract the entire organic fraction from the bone. In contrast, the cats generally show no real interest in crushing bones or extracting whatever nourishment is to be found in their marrow cavities, even though the larger cats are big and powerful enough to do considerable damage to bones. (Lions and tigers held in captivity may chew up much of the end of a large bone from a cow, but this activity may result as much from boredom as from any real desire to consume the bone.) It is however true that wild cheetahs have been observed to consume significant quantities of bones of very small prey, and to eat ribs and parts of the vertebrae of larger animals, so that capabilities and perhaps nutritional imperatives should not be overlooked.

If we examine the cheek teeth of a typical cat, whether it be a lion or a domestic tabby, we see that by far the most important feature is the scissorslike arrangement of the upper and lower carnassials, the meat-slicers. This arrangement is enhanced by the fact that the articulation for the mandible or lower jaw, the hinge, is in line with the intersection between the carnassials, just as in a pair of scissors. The other premolar teeth, although by no means unimportant to the animal, are relatively less significant. When we examine the cheek teeth of a saber-tooth such as *Homotherium latidens*, a species fairly common in Europe around 1.0 Ma ago, we

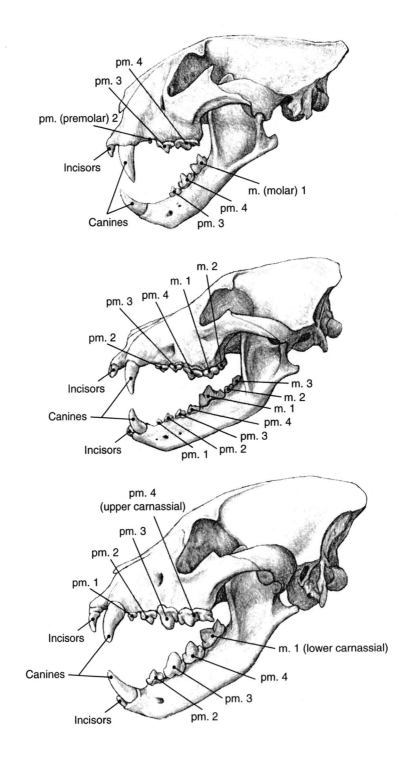

pm. 4
pm. 3
pm. (premolar) 2
Incisors
Canines
m. (molar) 1
pm. 4
pm. 3

m. 2
m. 1
pm. 3 pm. 4
pm. 2
Incisors
m. 3
m. 2
m. 1
Canines
pm. 4
Incisors
pm. 3
pm. 1 pm. 2
pm. 2

pm. 4
(upper carnassial)
pm. 3
pm. 2
pm. 1
Incisors
m. 1 (lower carnassial)
Canines
pm. 4
pm. 3
Incisors
pm. 2

see an even greater specialization in slicing, with the anterior cheek teeth much reduced in size.

The dogs have carnassials too, but they are only part of a dental armory that is augmented by many more premolars in front of the carnassials and by the crushing molars behind them. The dog is therefore a generalist when it comes to food-processing ability. In the hyenas the specialization is in almost entirely the opposite direction to the cats, with the development of huge, conical, bone-cracking teeth. Spotted hyenas in Africa today are capable of eating the entire carcass of a zebra, bones and all, but even they retain the carnassials to permit them to slice meat and other softer tissues.

Cats, then, are essentially flesh eaters: they eat primarily the meatier and softer parts of a carcass. But the mouth of a cat has a second notable feature in the size of its canine teeth. Those of the domestic cat are small but needle-sharp; those of a lion or tiger are massive conical structures—especially the upper ones, in which the unworn crown (the part that stands

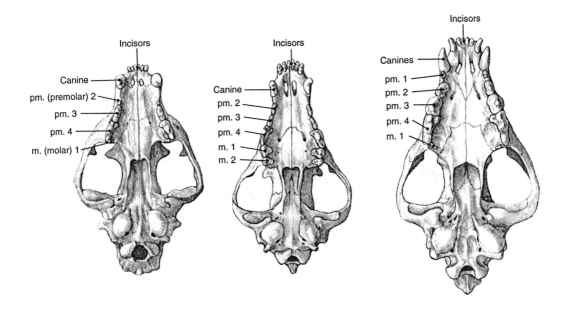

FIGURE 4.19 *Comparative views of skulls, mandibles, and teeth of cat, dog, and hyena*
The animals illustrated are a leopard, *Panthera pardus* (*top and left*), a wolf, *Canis lupus* (*center*), and a spotted hyena, *Crocuta crocuta*. Notice how the skull of the wolf is relatively long, and how it has many more teeth than either the cat or the hyena. It achieves this increase by retaining more of the basic mammalian dentition, with more molars and the full complement of four premolars.

The hyena has very large, bone-smashing premolar teeth, while the molars of the dog perform an important bone-breaking function. In contrast, the cheek teeth of the cat are almost entirely devoted to slicing.

above the gum) may be up to 8 cm long in a large male. Such teeth are not merely for show, but are used extensively for subduing and killing prey once it is firmly held by the sharp claws; they may also be employed in fights between rivals. In older animals it is quite common to find one or more of the canines broken, or worn down to a smooth stump of variable length following a presumed break. Many such breaks are doubtless produced during prey capture, either by a struggling animal fighting to set itself free or, as in some cases that have been observed, when the cat is kicked in the face by a zebra, a large antelope, or even a giraffe (figure 4.20). The canines also function, together with the incisors that lie between them in both upper and lower jaws, to help tear food off a carcass, and this action may lead to further wear and damage if bone is struck.

The saber-toothed cats carry the development of both the upper canines and the carnassials to extremes (but with important differences to be seen between the various lineages). However, it is as well to bear in mind, as we have already seen, that the saber-tooths were not unique, since at least three other groups among the mammalian meat eaters have developed their upper canines. Moreover, among our own order, the Primates, the upper canines of male baboons are fearsome-looking teeth, used extensively in display.

FIGURE 4.20 Nimravides *being kicked by a horse*
This reconstruction of events is based upon the pattern of damage seen in a skull and mandible from Kansas belonging to a particularly unfortunate individual. At some point in its life the animal broke an upper canine, but the heavy wear on the stump of the tooth shows that it survived this incident. Having survived for so long the cat was then unlucky enough to suffer a broken mandible, and since the fracture shows no sign of healing it is likely that the animal died soon afterward. One very likely cause for such injuries is the kick of a horse (as depicted here) or of some other large ungulate, often the last desperate defense of an animal about to be caught. Such accidents happen to African lions today when they hunt zebras or are unwise enough to tackle buffalo, and the cat often dies of starvation if the injury is sufficiently severe.

Among the saber-toothed felids the upper canines (but rarely the lower ones—except in some of the members of the extinct genus *Paramachairodus* and the living clouded leopard, *Neofelis nebulosa,* as seen here in figure 4.21) are usually elongated out from the jaw and flattened into bladelike shapes—hence the references to them in the literature as "saber"- or "dirk"-like. Such teeth often protrude below the lower jaw, or mandible, and in some cases are matched by a flange at the front of the mandible. The carnassials, both upper and lower, may become longer and narrower, like elongated scissors. In many cases the teeth between the canines and the carnassials are lost, become reduced in size, or develop their own bladelike characters, and in many species all three courses of evolution have been followed to some extent.

The cheek teeth of the saber-tooths therefore show a specialization toward slicing, one that is even further away from the limited bone-crushing abilities of the living cats. Such teeth must have been used with care if damage was to be avoided—and so too must the canines, in view of their length and slenderness. Numerous paleontologists over the years have therefore argued that subduing prey and killing it must have involved techniques that differed somewhat from those employed by the living cats. Many ideas have been proposed to explain the use of these enormously elongated teeth, with varying degrees of plausibility, but it is clear that they could *not* simply have been sunk deep into the neck of an animal in the process of leaping onto it and wrestling it to the ground, as many depictions have suggested: almost the slightest movement on the part of the struggling prey would have threatened breakage in such circumstances.

Strangely, the stabbing explanation in its various forms has tended to exclude the lower jaw (the mandible) from any mechanical calculations, simply carrying to extremes the analogy of the upper canines with a human wielding a steel knife (figure 4.22). Quite apart from the obvious risk of damage to the teeth, we see three main problems with this explanation. First, the machairodont canine is much more blunt than a steel knife. Second, driving such teeth deeply into the flesh of a victim in order to kill it during capture appears to us to require an excessive amount of force. Even if the necessary force could be generated through a combination of momentum and stabbing movements of the head, the risk of damage to the teeth through hitting bone or simple torsion during insertion would have been significant. Third, whatever the angle of attack chosen, it is difficult to see the mandible as other than an impediment to effective stabbing.

As an alternative to their use as a means of stabbing prey to death it has been suggested that the canines would have been mainly used to slice into the body of the prey in some manner, to make the flesh more accessible, once the animal had been killed. The first problem with that explanation is that it requires the prior assumption that the machairodonts existed largely as scavengers, since the method of killing prey is not considered.

FIGURE 4.22 *Suggested parallels between a saber-toothed cat and a human wielding a knife*
The "stabbing theory" explanation for the use of the saber-tooth canines implies parallels with a human hand wielding a knife. But a steel blade is quite different from a machairodont canine, which is too blunt and too fragile to have functioned in this manner with any consistency.

FIGURE 4.21 Paramachairodus ogygia (top) *with flehmen gesture compared with a modern clouded leopard,* Neofelis nebulosa
The flehmen grimace bares the teeth of the extinct cat, allowing us to see that the lower canines are quite elongated in this species, an unusual feature even in animals with excessively elongated upper canines. In this respect it resembles the living clouded leopard, *Neofelis nebulosa*, and the similarity led early workers to think that the two genera were closely related. But the upper canines of the fossil cat are more flattened, and in this and other skull characters it shows many more machairodont affinities.

Although virtually all living predators do scavenge to some extent, it is difficult to imagine that the machairodonts were generally able to adopt such a strategy in the face of competition for carcasses from the other large carnivores with which they coexisted. The second problem is that the wear on the canines, if they were habitually used in this way during feeding, would be greater than that generally observed: in many cases the canines are actually much less worn than the other teeth of individuals. It has also been pointed out that the sockets of the canines do not show the pattern of bony reinforcement about the opening that would be expected from such habitual use. As a living tissue, bone responds to stresses placed upon it, often by building more thickly where needed—in this case, where the bone emerges from the jaw. Instead, the sockets of the canines exhibit reinforcement toward the base, a pattern consistent with some kind of piercing action even if it fell short of a strongly stabbing motion.

What we seem to be left with as a general explanation is one put forward several years ago by the American paleontologist William Akersten, who suggested that a "shearing bite" to the abdomen of a temporarily immobilized prey might produce a gaping wound that would lead to significant blood loss and probably shock (figure 4.23). Akersten's model saw the enlarged flange at the front of the mandible as an anchor point against which the head depressor muscles might force the upper canines, producing a pierced rather than a stabbed wound. In the case of *Smilodon* in particular, he suggested that group action against juveniles of large species such as mammoths might even have been possible, with the group perhaps retreating for some time after the initial attack and waiting for the victim to die before returning to feed. Leyhausen describes how domestic cats may leave freshly dead prey for some time, and perhaps one could extend such behavior to machairodont kill techniques. But larger carnivores seem somewhat reluctant to lose contact with their prey once it is caught, and tend to retreat only if directly attacked, and we think it unlikely that a group of *Smilodon* would risk the seizure of their meal by other predators in the vicinity in such a way.

However, we do find Akersten's general model rather plausible, since it explains many features of the overall morphology of the machairodonts while avoiding the objections to other interpretations that rely in some way

FIGURE 4.23 *A sketch of William Akersten's ideas*
The drawings show the scimitar-toothed cat *Homotherium* applying a shearing bite to its prey. In the *first drawing*, the upper canines begin to pierce the skin. In the *second drawing*, the mandible gives support as the head is depressed, and the upper canines penetrate farther while the lower canines and incisors add to the wound. In the *third drawing*, the fold of skin and flesh is pulled back with the jaws closed, causing considerable loss of blood and perhaps tearing off a whole chunk.

or other on the seemingly obvious analogy with an aggressive, stabbing mode of action. We also suggest that a shearing bite could be employed to the throat region of the neck of the prey, but only once the animal had been brought down and held virtually immobile, not while it was still standing (plate 9). Here the great strength that we see in the machairodonts, particularly in the forequarters of *Megantereon,* would have been important. As the illustrations in figure 4.24 show, once the prey is held still, it is possible to bite deeply into the throat, doing massive and, more importantly, rapid damage to the windpipe and major blood vessels.

Whatever the true method of killing, the large size of the canines would also have meant that the method of getting food from a carcass differed from that employed by living cats, since they could not have been used in conjunction with the incisors to pull meat and skin away from the bones: they are simply too long. There is a further difference, however, exemplified in figure 4.25. In the living cats, such as the lion, tiger, and leopard, the six incisors in each upper and lower jaw are deployed as an essentially straight row of teeth between the left and right canines. The upper incisors are set somewhat forward of the line between the upper canines because the latter teeth actually sit somewhat behind and outside the lower canines when the mouth is closed, producing a gap into which the lower canine can fit. The central pair of incisors are the smallest, the two outside those are somewhat larger, and the two nearest the canines are largest of all. But even in a large lion the incisors are quite small teeth. In contrast, the incisors of many of the saber-toothed cats are often large, and the upper ones in particular are deployed in an arc that sets them well in front of the upper canines. This morphology is seen especially in the case of *Homotherium,* in which a view from the side shows the canine falling almost midway along the total extent of the dentition, and only slightly less so in *Smilodon.* In the case of *Megantereon* the protrusion of the upper incisor arc is also quite marked; and even in *Dinofelis,* with flattened but relatively short canines, the upper incisors are large. In other words, the incisor teeth in the saber-toothed species are developed and positioned in a way that overcomes the

....

FIGURE 4.24 Megantereon cultridens *biting through the throat of a horse*
Could the saber-toothed cats have bitten into the throat of a large ungulate without risk of damage to their teeth? The illustration shows a schematic section through the neck of a typical horse, with the skull of *Megantereon* biting at the throat. Notice how the vertebrae of the horse are arranged toward the back of the neck, and how near the surface of the throat are the windpipe and major blood vessels. With the animal held immobile by strong forequarters, even relatively superficial slashes into the neck would produce considerable blood loss and induce shock, and choking off the air supply would be relatively easy. Such a technique would avoid the need for the violent and rather inaccurate stabbing implied by some older ideas about how machairodonts dealt with their prey.

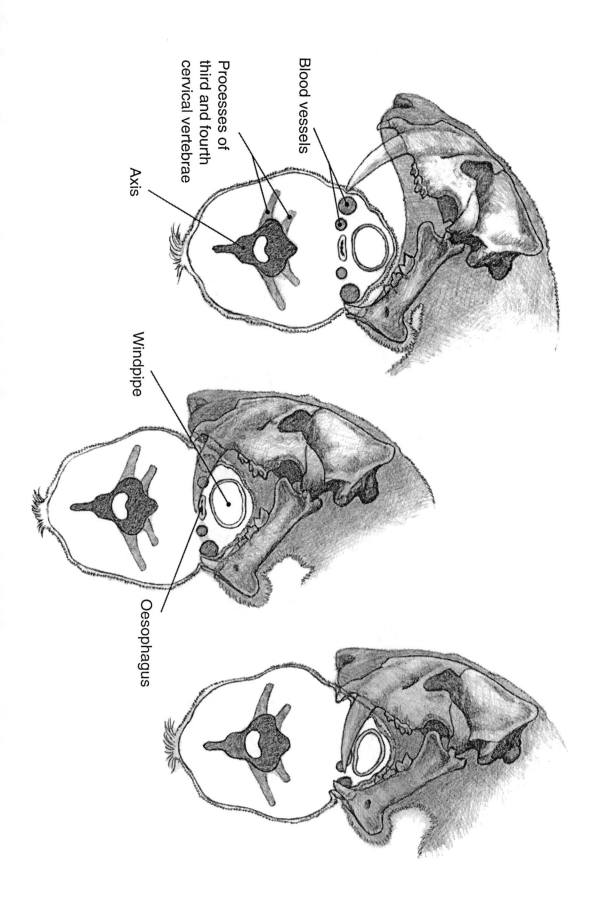

Blood vessels

Processes of
third and fourth
cervical vertebrae

Axis

Windpipe

Oesophagus

problem posed by the canines, and that permits the animal to get at and remove flesh from the bones.

In many ways the arrangements of the incisors in the saber-toothed cats are therefore more like those of a wolf or a hyena, as seen in figure 4.19. Such a positioning might have allowed the animals to grasp the skin of their prey and to drag it backward, something that the large living cats

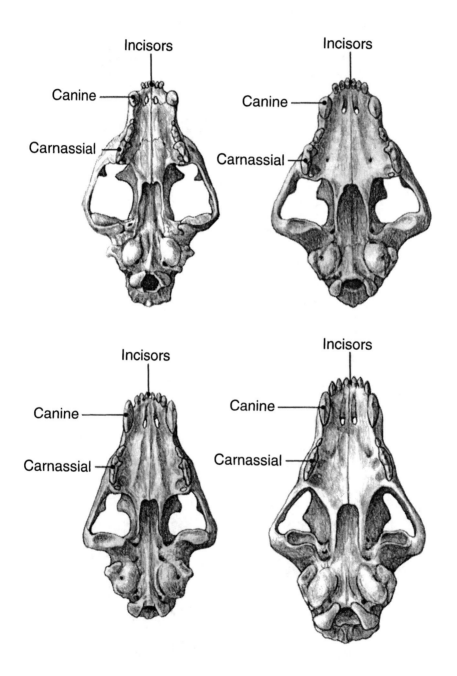

may do—although this is perhaps not as efficient as moving a carcass by straddling it and walking forward with at least part of the body held off the ground.

SKULL MORPHOLOGY

We can now return to the question of skull morphology. In comparison with those of the Felinae, the skulls of the Machairodontinae differ in a number of key characteristics. In particular, we may list the following, some of which are shown in figure 4.26:

1. the face is rotated upward relative to the braincase;

2. the articulation of the mandible or lower jaw is moved downward;

3. the distance between the articulation and the carnassial is reduced;

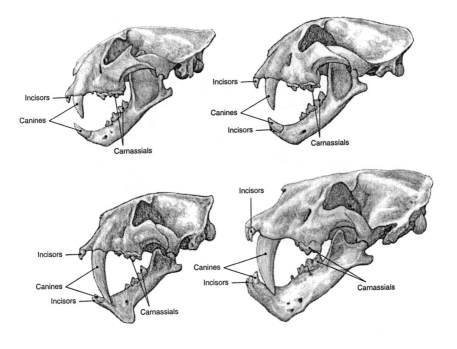

FIGURE 4.25 *Side and palatal views of leopard* (top left), Dinofelis (top right), Megantereon (bottom left), *and* Homotherium (bottom right) *to illustrate the disposition of the incisors and canines*
Note the large size and forward placement in an arc of the incisors in the fossil species in comparison with the leopard.

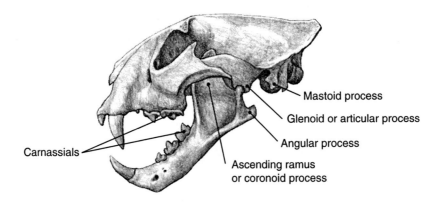

Mastoid process

Glenoid or articular process

Angular process

Carnassials

Ascending ramus
or coronoid process

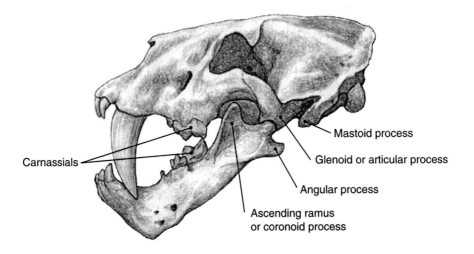

Mastoid process

Glenoid or articular process

Angular process

Carnassials

Ascending ramus
or coronoid process

FIGURE 4.26 *Comparison between feline and machairodontine skulls*
In these lateral views of the skulls of a leopard (top) and *Homotherium* (bottom)
we can see at a glance most of the major differences discussed in the text.

4. the vertical portion of the mandible, the ascending ramus to which the temporalis muscle attaches, is reduced;

5. the temporalis muscle is oriented more vertically;

6. the angular process of the mandible below the articulation is flared outward.

Of course differences still existed between the various machairodonts, and not only between genera but also between species within genera. The skulls of *Smilodon populator* and *S. fatalis*, for example, differ quite significantly, as has been pointed out by the Scandinavian paleontologists Björn Kurtén and Lars Werdelin. They suggested that the South American species may have carried its head lower because of differences in the orientation of muscle attachments at the rear of the skull, with the result that the facial part was rotated upward even further in compensation.

All the features of the machairodont skull listed above are related to increasing the maximum gape, which was approximately 90–95° in the machairodonts (in contrast to 65–70° in living cats), to increasing the strength of bite at the carnassial, and to overcoming conflicts between the two requirements. By increasing the gape to such a significant extent, the saber-tooths achieved approximately the same clearance between upper and lower canine tips as that achieved by modern cats. They were therefore not disadvantaged by the increased size of the canines, at least so far as their ability to gape was concerned.

Despite the clear efficiency of these features and the interrelated nature of the cranial adaptations, the saber-toothed cats have provoked the strangest ideas from evolutionary commentators. These have included the suggestion that the large canines meant that these cats attacked with their mouths closed and then were unable to open them in order to eat! Other arguments for extinction have included the possibility that the tips of the upper and lower canines would become interlocked, jamming the jaws half open and once more resulting in death by starvation. But these features are paralleled to a considerable degree in the skulls of other saber-toothed species among the Nimravidae, marsupials, and creodonts, pointing to the strong selective pressures and indicating that the possession of saber-tooths was far from being the bizarre and evolutionarily short-sighted development that it is often claimed to have been. It seems likely that there are only so many evolutionary outcomes to selection for a carnivorous way of life, given the need to catch, kill, and eat prey, and the saber-toothed pathway is one that has been followed now on several occasions. It is also worth stressing that the Nimravidae further paralleled the Felidae in the development of retractable claws in conjunction with the development of saber-toothed dentitions. Such claws are themselves of major importance to the functioning of the living cats, and should be seen as part of an overall adaptation for catching and dispatching prey.

Claws

All the members of the Carnivora have claws. They are usually sharp, often curved, and sometimes impressively long. They grow from the end of the third phalanges (the phalanges are the bones that make up what we might call the fingers and toes of the paws), just like our own fingernails, and they are made of the same material, keratin. The third phalanx of the cat has been modified to form a core around which the keratinized sheath of the claw grows in a sharply curved arc. Such claws can function as weapons, prey-capturing aids, and grooming tools. But, again like our own fingernails, they are made of relatively soft material, and will wear and break. The effect of wear on claws, and its absence, can be easily seen by comparing the feet of a dog unused to walking on hard ground with those of one regularly exercised over paved pathways or other hard surfaces.

One of the most obvious features of the cat family in comparison with the dogs, however, is the possession of retractable claws that can be deployed at will. This ability, shared only with a few members of the family Viverridae and with the extinct Nimravidae, preserves the sharpness and length of the claws during everyday activities. But how is it achieved?

As the diagrams in figures 4.27 and 4.28 show, the secret lies in the shape of the phalanges. The claw itself grows as a horny sheath from the third phalanx. The third phalanx in turn fits onto the end of the second phalanx, but the end of the second phalanx has a peculiar feature: the articulation is offset to one side, allowing the third phalanx to retract upward and backward to lie in part alongside the second. This movement is enhanced by the angled nature of the articulation on the third phalanx. It is as though you could take the end joints of your own fingers and toes and bend them backward and somewhat sideways, toward the knuckles, instead of inward toward the palm as is usually the case.

This retracted position is in fact the normal resting place of the cat's third phalanx, achieved by the action of elastic ligaments that pull the phalanx (and with it the claw) back so that in relaxation the sharp tip is drawn back into the fur of the paw. If you examine the paw of a domestic cat, which operates on precisely this general principle, you will find that the tips of the claws are perhaps hard to see but can be easily felt when the cat is at rest.

The claws are deployed by swiveling the third phalanx round on the offset articulation of the second phalanx, throwing the claw forward. Again, if you examine the paw of a domestic cat you may, with care, be able to get the animal to demonstrate this ability. Notice how as the claws emerge the whole paw seems to expand, rather as your own hand does if you hold it out and extend and splay the fingers. In this position the paw makes a very

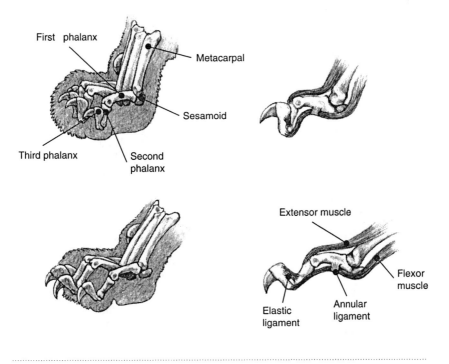

FIGURE 4.27 *Claw retraction*

The mechanism of claw retraction and deployment is an elegant example of natural design in a bodily part required to perform a number of functions.

Above left: The articulated bones of the left forefoot, or manus, of a lion with retracted claws. The third phalanges are folded against the left side of each second phalanx.

Above right: A side view of the central digit shows the muscles and tendons in a relaxed position. The third phalanx is kept in position against the second phalanx by an elastic ligament.

Below left: The articulated bones of the same foot with extended claws.

Below right: As the digits are extended, the tendon of the flexor muscle pulls on the lower edge of the third phalanx so that the claw turns and points downward.

effective device for defense or offense. In prey capture, the extended claws can be dug into the skin of an animal and will enable the cat to cling on; if it chooses to, the cat can close the extended claws much as we would close our own fingers around an object, further digging the points into the skin of its quarry and making escape very difficult. This ability also serves to help the cat climb trees, but it can make descent somewhat hazardous because the deployment works better for ascent and provides little grip on the return journey. This explains why your domestic cat can virtually run up a tree but may have to be rescued with a ladder.

The cheetah is often said to be unique among the felids in having claws that are somewhat less retractable. Indeed, the genus name *Acinonyx* (from

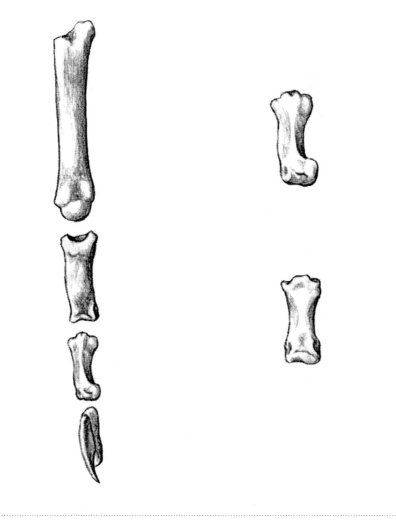

FIGURE 4.28 *A dorsal view (from in front) of the bones of the central digit of a cat*
In the second phalanx the distal half and the facet for articulation with the third phalanx are offset to the side, which means that the third phalanx can fold back to lie alongside the second. This asymmetry is even more evident if we compare the second phalanx of a cat (*above right*) with that of a dog (*below right*).

Greek *akineo*, no movement, and *onyx*, claw) refers to that very point. However, it is not true that its claws are nonretractable, and it would seem that wear and tear upon them during the chase, together with some differences in the soft-tissue morphology in comparison with other cats, have been largely responsible for the confusion. The cheetah's claws are not partially covered by a sheath of skin during retraction, making it look as though they are still somewhat deployed. The morphology of the second and third phalanges is virtually identical to that of other cats. The claws

themselves appear rather straight, adding to the impression that they protrude from the fur of the paw in an unretracted manner—but this, at least
in part, presumably reflects use, perhaps to enhance the grip when turning
quickly. The animal uses its claws to bring down and hold its prey in its own
unique manner, as we shall see when we discuss its hunting techniques in
the next chapter, and it could not do this if they were not retractable and
protected from the worst of wear.

One further feature of the claws in the cats, shared with most of the
dogs, is the possession of a large dew claw, as it is often known, on the inside
of the front paw, set some distance above the others. This is the equivalent
of our own thumb. Because it does not come into contact with the ground,
it is invariably long and sharp, but it is not functionless. The dew claw forms
an important part of the prey-grasping technique employed by many of the
cats, and it is especially well developed in the cheetah where it is large and
hooklike and very much more like the claws of other felids.

The development of fully retractable claws in conjunction with large
canines among the Nimravidae (figure 4.29) points to the need to hold and
immobilize prey if the canines are to be used without risk of breakage. It is
no coincidence that the saber-toothed species among the powerful-looking
Nimravidae are among the most extreme in the development of the upper
canines, and that the saber-toothed species among the Machairodontinae
are often robustly built animals. Even among the conical-toothed true cats
of the Felinae, the advantages of holding prey firmly while dispatching it
with a characteristic bite to the neck are abundantly clear, as we shall see in
the next chapter. It seems likely that the parallels between the Nimravidae
and the Felidae, for all that the two groups appear to deserve separation at
the family level, indicate a relatively close relationship between the two in
the evolution of the Carnivora.

RECONSTRUCTION: THE FRAMEWORK

Figure 4.30 illustrates some of the principles of reconstruction from fossil
evidence of two morphologically similar species. The vertebrate body can
be seen as a system of levers, where the bones give support to the action of
muscles and are in turn moved by them. Most muscles are attached, via tendons, to the bones (some, such as the heart and the smooth muscle of the
stomach, gut, and blood vessels, are of course not), and the morphology of
the bone at the site of insertion of the tendon often betrays this fact, sometimes by responding to the stress placed on it by the action of the muscle
with a change in shape. Ridges and crests may be developed, and the more
or less rugose (roughened) surface of the bone at the site of insertion can
therefore provide clues to the strength of the muscle. With knowledge of
the muscular anatomy of living species we can therefore study the bones of
a fossil taxon and reach a number of conclusions about the general bodily

FIGURE 4.29 *A sketch of* Barbourofelis fricki *scratching a tree*
The illustration shows the basic similarity in body plan between felids and nimravids, and highlights the possible parallels in claw use and maintenance behavior
between the two families: anyone who has ever kept a domestic cat will be familiar
with this behavior pattern. It seems that the scratching does not really sharpen the
claws; rather, it removes old claw sheaths and actually keeps the claws reasonably
worn. Captive cats and viverrids that are not provided with a scratching surface
often get overgrown claws, which may sink into the flesh of the footpad and crip-
ple the animal.

appearance of the living animal and the manner in which its system of
levers worked.

In addition to the information to be gleaned from the normal bony
anatomy of our fossils, we have a further avenue of inquiry in the form of
abnormal, or pathological, conditions of the skeleton. A secondary growth
of bone in the region of a muscle attachment site may mean that the muscle was repeatedly torn by stress. As we pointed out in chapter 3, such
growths have been found on the bones of *Smilodon* from Rancho La Brea,
for example, especially in the insertion areas of the deltoid muscle on the
humerus (figure 4.31). The frequency of such pathologies suggests that
they were a fairly common problem for the animal, perhaps the result of
stress during lateral movements of the limbs such as would typically occur

when subduing large and struggling prey. If that is true, then we gain an important insight into the life of *Smilodon* and into the relative importance of adaptations. We are easily led into thinking the robust nimravid dirk-tooths to be primitive creatures lacking the advantages of the cursorial adaptations enjoyed by modern felids. But the advantages conferred by their short limb bones and prominent crests for muscle attachment were considerable, and *Smilodon*, powerful as it was, may have paid a very real price for the course that its own evolution followed.

On the other hand, the Oligocene nimravid dirk-tooth *Hoplophoneus* often shows pathological conditions on its lumbar vertebrae, derived no doubt from stress. In this respect, at least, *Smilodon* may have had the advantage of a shorter and therefore more powerful back, although stress-induced lesions are still common in the Rancho La Brea sample. Interestingly, a pathological bony growth similar to those seen in *Smilodon* is found on the humerus of another machairodont, *Homotherium latidens* from Senèze in France (figure 4.31). This suggests that, despite its particular cursorial traits, *Homotherium* was still faced with the problem of bringing down large prey with the force of its front limbs.

A less spectacular but no less significant example of how muscle attachments relate to function in fossil species is seen in the ankle and foot of *Pseudaelurus* and other Miocene cats (figure 4.32). In the calcaneum of these primitive animals there is a clear attachment for a muscle called the quadratus plantae, which unites the calcaneum with the toes via the sole of the foot. This attachment is all but invisible in the calcaneum of living cats (and Pleistocene saber-tooths), and the muscle itself is very small. The size of the attachment in primitive saber-tooths and the "feline" *Pseudaelurus* may be explained by the retention of an emphasized grasping function, since the muscle flexes the toes relative to the ankle.

Interestingly, the quadratus plantae muscle is well developed in the arboreal fossa, *Cryptoprocta fossa*, of Madagascar, which shows other adaptations similar to those of the earliest cats. During subsequent evolution, as terrestrial locomotion became dominant in all cats, the quadratus plantae became less important and another muscle, the flexor digitoris longus, took over: this flexes the toes relative to the tibia, a propulsive rather than a grasping function. Some support for this interpretation may be seen in the imperfect digitigrady revealed in some recently discovered felid footprints in lower Miocene deposits of Salinas de Añana in Spain (discussed further below).

BRINGING FOSSIL CATS TO LIFE

To reconstruct the life appearance of a cat, we start by drawing the assembled skeleton in a life pose. Most of the species depicted in this book are represented by a reasonable number of bones, or even complete skeletons, but much of the material was unpublished. A rather arduous search plus an element of luck have therefore gone into the collection of reconstructions offered here. In some instances the material is abundant but without a complete skeleton of any one animal, and in such situations we have had to scale the bones of differently sized individuals to fit, relying on common sense; the caption to each reconstruction gives some information on the material on which it was based.

With the skeleton assembled we put the deep muscles in place first (figure 4.33). Because not all of these leave clear markings on the bone we have had to employ comparisons with related living taxa, but the result is not likely to differ significantly from the true condition. In any event, many of the muscles that are often termed "superficial," shown here in the second drawing, leave clear markings, and the image of the animal that is being built up at this stage is firmly based upon skeletal information.

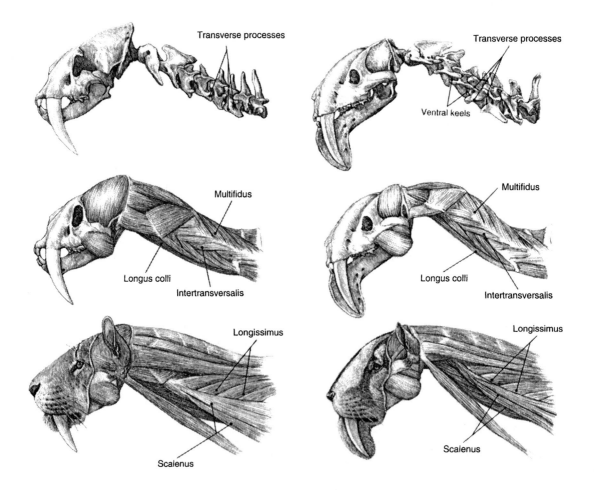

Special attention must be given to the restoration of the head, since it embodies so much of the "personality" of the animal. The shape and proportions of the skull are essential for definition of the living appearance of the head, and we have always remained faithful to those basic features. But at times it can be difficult to get the information needed for reconstruction. In scientific papers only orthogonal views (lateral, frontal, superior, and so on) are usually given, so in some cases we sculpted a model to match those views and then used the model to draw the skull from other angles. Sometimes the fossil material has been crushed or otherwise distorted by the sediments, and in such cases we have had to reconstruct the shape of the skull carefully. Of course there are now computerized ways of doing this that are becoming more and more common, but we have not had the benefit of such assistance for this project.

Once the skull is drawn, we first put in place the masticatory muscles: these define the main volumes of the head, and their attachment surfaces, fortunately, are mostly marked by clear features of the skull. Other parts, such as the length of the cartilaginous nose, the position and shape of the external ears, and the lips, are more difficult to determine from bone shape, and yet they are important in any effort to define how the animal looked. Here opinions differ. It has been proposed, for example, that felid saber-tooths may have looked very different from living felines not only in

FIGURE 4.30 *Head and neck of* Smilodon (left) *and* Thylacosmilus *showing deep muscles* These two unrelated saber-toothed animals independently developed necks that are longer than in their non-saber-toothed relatives. In both animals the vertebrae of the neck exhibit enlarged processes for muscle attachment, and this has usually been considered a reflection of a very strong head-depressing musculature, necessary for a stabbing action. A careful examination of the vertebrae shows that while the insertions for head-depressing muscles such as the scalenes are indeed well developed, the attachments for other muscles that turn the head and neck to one side or pull it up are equally (if not more) important.

The two animals inherited somewhat different vertebrae from their ancestors, so that the precise result is different in each case. *Thylacosmilus* has prominent ventral keels on the vertebrae, and a powerful depressor of the neck and head, the longus colli, inserts there. But the ancestors of the Felidae had lost those ventral keels, and in *Smilodon* other head-depressing muscles, such as the scalenes, are more emphasized. These latter muscles insert at the lower part of the transverse or lateral processes of the vertebrae, which are very developed in both animals. However, the shape of those transverse processes implies that other muscles were also involved. The intertransversalis (which turns the neck laterally) and the longissimus colli (which extends the neck and turns it) have their attachments there, and they probably had an enhanced function in both saber-tooths. This may imply that, if the animals killed with a shearing bite, a long neck and a quick, precise turning of the head were important for reaching particular parts of the body of the prey and positioning the teeth for the rather careful bite. During the bite itself, head-depressing muscles were indeed important, while for the final, backward pull the extensor muscles would have entered into play.

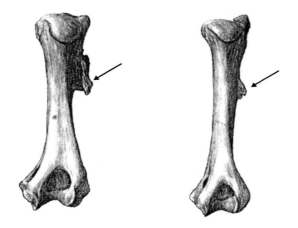

FIGURE 4.31 *Bony growths in humeri of* Smilodon (left) *and* Homotherium
Pathological bony growths of this kind (*arrowed*) are often seen on bones where
there has been a pattern of recurrent stress producing a tear.

terms of their visible dentition but also in precisely those features that must
be reconstructed by inference. This view has led the American paleontologist George Miller to suggest three major points of difference for the
head of *Smilodon fatalis* in particular, with clear implications for other
species (figure 4.34). The first concerns the relative position of the ears,
which he has argued would be placed lower on the skull than in modern
cats because the high sagittal crest would raise the top of the head. The second concerns the nostrils, which he believes would be in a retracted position based on the short nasal bones and would give a bulldoglike effect
exaggerated by the prognathous incisor battery. The third point is the lips,
which in his view would have had to be longer and more doglike in their
appearance to permit the large gape needed for the upper canines to clear
the lower ones and thus enable food to be got into the side of the mouth
and cut by the carnassials.

These are important points, because they imply a very unusual appearance for those animals. In our reconstruction we have taken a different
view, but one that we believe is justified. Firstly, the placement of the ears is
not strictly tied to the position of the opening in the skull, which does
indeed appear relatively low in *Smilodon*, but is more closely tied to the functional aspects of hearing (as mentioned earlier in the chapter). Thus, in the
living serval cat the external ears nearly touch each other in the center of
the skull, while in the ocelot they are markedly more laterally placed. When
the skulls alone are considered such differences are less apparent, because
the soft tissues of the ear make up the major difference. As to the development of a crest, and its bearing on the catlike nature of the head, this feature is variously present in living cats such as the Spanish lynx (absent), the

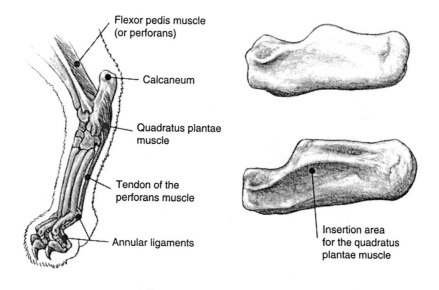

Flexor pedis muscle
(or perforans)

Calcaneum

Quadratus plantae
muscle

Tendon of the
perforans muscle

Annular ligaments

Insertion area
for the quadratus
plantae muscle

FIGURE 4.32 *Calcaneus of* Pseudaelurus *and a modern cat*
The illustration shows the differences in the muscle attachments of the bone,
from which we can infer a greater grasping ability in the foot of *Pseudaelurus*. The
important difference stems from the presence of the quadratus plantae muscle in
Pseudaelurus. As shown in the drawing on the left, this muscle (also known as the
accessory flexor) arises from the side of the calcaneus and joins the tendon of the
flexor of the digits in running across the sole of the foot. When the muscle contracts it permits the toes to flex without the involvement of the normal toe-flexing
muscle, which originates on the tibia. More-cursorial animals need to eliminate
weight from the feet, hence the reduction of this muscle in modern cats. The calcaneus of *Pseudaelurus* (*below right*) shows a ridge (surrounded by grooves) where
the quadratus plantae originated. These features are absent in the calcaneus of a
modern cat (*above right*), but are present in other, noncursorial living carnivores.

cheetah (modestly developed), and the jaguar (pronounced). All these
species certainly differ in appearance, yet all are perfectly catlike, and we
believe that so too were the saber-tooths.

So far as the nose is concerned, there is again variation in the size of the
nasal bones of living felines, and in the largest members of the genus
Panthera the incisor battery is proportionally farther away from the tips of
the nasals than in smaller cats. Lions have the most retracted nasals of the
pantherines, much more so than in tigers, but the position of the rhinarium
(external nose) is not retracted in the lion and therefore the essentially catlike appearance of the living head is maintained. In all cases the nasal cartilage spans the distance, and it seems to us that the length of the cartilaginous nose in felids is always enough for the rhinarium to be in a similar position above the incisor battery. The retracted nasals of the felid saber-tooths

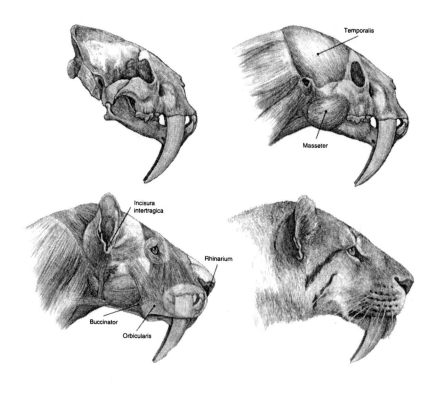

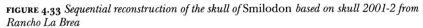

FIGURE 4.33 *Sequential reconstruction of the skull of* Smilodon *based on skull 2001-2 from Rancho La Brea*

In this sequential reconstruction, and after positioning the masticatory muscles, we show the likely position of facial muscles, mouth opening, nose, and ears.

The orbicularis is an important muscle in carnivores, making up much of the wall of the mouth. In the modern lion this muscle extends about 5 cm from the lip line, and we have supposed a similar extension in the lion-sized *Smilodon*. This positioning helps us define the extension of the lip line, since the orbicularis is always placed ahead of the anterior end of the masseter, and in fact there should be a small space between the two muscles for the fibers of the buccinator, the other main contributor to the formation of the wall of the mouth. The nerve bed for the mystacial vibrissae (the whiskers) has been drawn here as a transparent surface, allowing us to see the anterior teeth through it. The rhinarium is placed above the upper incisor row, as is normal in living carnivores. The incisura intertragica (the lowermost part of the external ear opening) is placed slightly above the level of the auditory meatus (the hole of the ear), a condition seen in a dissected cheetah and apparently present in other large cats as judged from comparisons of photographs and skull profiles. The ear is depicted as relatively smaller than in lions, since *Smilodon fatalis* lived in a more temperate environment, but of course this remains hypothetical.

seem to be at the end of the continuum from smaller to larger cats, rather than something unique. To our knowledge, no wild carnivore exhibits the bulldoglike profile proposed by Miller for *Smilodon fatalis*, no matter how prognathous their incisors relative to the canines. Indeed, in the case of the bulldog the effect is produced by the retraction of the whole mid-face, along with the upper incisors, so that the lower incisors protrude and no longer occlude properly—an aberrant and highly maladaptive situation not seen in the machairodonts, and hard to envisage in natural circumstances. In our reconstructions we have therefore fixed the position of the rhinarium above and slightly ahead of the upper incisors, as is normal in carnivores.

The suggestion that the wide gape of the machairodonts as exemplified by *Smilodon* would exceed the elastic properties of the mouth, lips, and cheek muscles of a normal cat and thus would necessitate a larger mouth and longer lips may underestimate the elasticity of the tissues. We have observed lions yawning, and once the mouth is open at fully 70 degrees the

FIGURE 4.34 *An alternative suggestion for the head of* Smilodon
Did the living *Smilodon* look like this? The rather strange appearance of this head results from following the proposals of the American paleontologist George Miller. See the text for a discussion of these arguments, and see figure 4.35.

animal makes a grimacing gesture that shows the elasticity of the mouth to be well in excess of the demands of the normal gape. If we take the gape of *Smilodon* to be about 30% more than that of a lion, then the elasticity of the mouth would probably mean that less than a 30% increase in lip length was needed to accommodate the gape—certainly less than Miller's reconstruction, which called for a lip line half as long again as that of a lion (figure 4.34). In any event, if we look at living members of the Carnivora we find that in all of them the lip line reaches back to a similar point relative to the carnassials; this is seen in dogs just as in cats. If dogs have a longer mouth opening it is because they have a longer muzzle: in other words, the lips could be seen as extending *forward* from the carnassials rather than backward. But there is one other, entirely practical reason for believing this position to be correct. If the lip line were to go any further back, then it would encounter the muscles that attach to the mandible and form the cheek; these muscles thus set the limit for the lip line. In our reconstructions we have therefore kept to this principle and placed the back end of the mouth near the normal position. With this mouth size, the flexibility of the lips allows food to be taken laterally with the carnassial shear, as shown in figure 5.28.

The question of whiskers we have already dealt with, but as a final point we should return more generally to the question of the color patterns of the hairs in reconstructions. The decision to show a particular species with a spotted or plain coat is deliberate and is based on function. Thus, we have tended to reconstruct forest-dwelling animals as spotted and plains dwellers as less marked—but there are no absolutely fixed rules, and examples to the contrary (such as cheetahs and pumas) warn us to be cautious. But in all cases the body color must be reasonably practical (plate 10); melanistic (black) forms, for example, will not be successful in open terrain and today are restricted to populations of leopard, jaguar, clouded leopard, and serval living in relatively dense cover.

Nearly all living cats have some degree of facial marking, which can be striking even in cats with plain bodies like the puma, caracal, and golden cats. This universal incidence of facial markings emphasizes expression and may serve to aid communication and minimize intraspecific aggression, a matter of some importance in animals so lethally armed. We think it makes sense to expect that, as felids at least as dangerously armed, the machairodonts would have exhibited just as clear and catlike markings on their faces.

RECONSTRUCTING MOVEMENT

If we put together some of the information summarized above, we can use it to reconstruct various patterns of movement in extinct species for which we have only skeletal evidence. We therefore offer a brief analysis of some of the locomotor features seen in the cats.

Running

It is clear that no cats, living or extinct, have been well adapted for sustained running. In this they contrast markedly with the dogs and hyenas. We can illustrate this point best by considering three reconstructed running sequences.

RUNNING SEQUENCE 1: *Hoplophoneus* **(Figure 4.36)**
What we may term the "primitive" design is seen in early felids and nimravids alike. It is exemplified by *Pseudaelurus* and, with some inference, by *Proailurus,* but it is seen to best effect in the much better known nimravid *Hoplophoneus.* The design consists of a long and somewhat flexible back allied to shortened legs and, in particular, to shortened metapodia (the long bones in the feet equivalent to the bones in our own palms). It is also seen very clearly and in more extreme form in the structure of living animals such as mongooses, and in many smaller mustelids, the family to which stoats and minks belong.

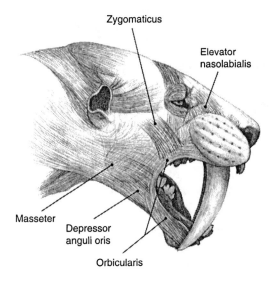

FIGURE 4.35 *Illustration of gape and the positioning of the lip line in saber-tooths*
In this illustration we show *Smilodon fatalis* with a grimacing gesture, with jaws open at about 35°. A combined contraction of several muscles of the face (the zygomaticus, the depressor anguli oris, and the elevator nasolabialis) pulls the fibers of the orbicularis from their resting position and bares the cheek teeth. To pull the orbicularis farther back would be pointless in terms of a carnassial bite, since the masseter muscles would get in the way of the food item. In the drawing, the mass of the masseter is insinuated under the fibers of the depressor anguli oris muscle.

Such a design seems well fitted to movement by the "half bound," a kind of gallop that uses a lot of energy but is more than adequate for short rushes because it permits quick acceleration. In this method of running, both hind legs leave and hit the ground almost simultaneously, while the forelimbs stride as in a conventional gallop. The technique takes advantage of the powerful extension of the back, which is expressed in a long phase of "extended flight." In contrast, there is only a short phase in which the limbs are gathered together beneath the body.

RUNNING SEQUENCE 2: *Megantereon* **(Figure 4.37)**
Modern big cats such as lions and tigers have longer legs and shorter backs than their Miocene ancestors, the pseudaelurines. This difference seems to reflect adaptations for improved locomotion on the ground. Observations from high-speed films show that while these animals may resort to the more primitive half bound of sequence 1 in the initial movements of a rush (presumably to gain acceleration from a standing start), they then change to a more classic gallop. The advanced saber-toothed cats of the Plio-Pleistocene had even shorter and more stiffened backs than the living pantherines, and would almost certainly not have been successful employers of the half bound. But the great strength of *Megantereon*, for example, suggests that such cats may have been ambush hunters, needing at least a few bounds to gain speed and close on their prey before it had chance to turn and flee.

RUNNING SEQUENCE 3: *Miracinonyx* **(Figure 4.38)**
Cheetahs often approach their prey with a confident trot, seemingly assured by their greater speed. Once they begin to run, the acceleration is rapid, as a result of their very particular morphology. The running cats of the *Acinonyx-Miracinonyx* morphological group at first sight resemble a greyhound dog in their overall body plan, with an elongation and thinning of the lower limb bones and a reduction in the musculature of the lower limbs and paws. Closer inspection shows, however, that they have a longer and more flexible back than dogs. The gallop employed by these cats is more energetically costly than in dogs, but there is a considerable gain in speed stemming from the long stride of the "suspended flight" phase interspersed with a longer "gathered" phase than that seen in any other cat.

Climbing

The anatomical features that enable big cats such as leopards to climb trees so successfully are closely related to their ability to bring down and hold struggling prey. This is not surprising, since most of what we consider to be adaptations fitting animals to their way of life turn out on inspection to be compromises between sometimes competing demands. Most adaptations are therefore less than perfect—we ourselves walk erect, with obvious advantages in freed hands, but we pay a penalty in back pain and associated

FIGURE 4.36 *Running sequence of* Hoplophoneus

FIGURE 4.37 *Running sequence of* Megantereon

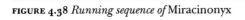

FIGURE 4.38 *Running sequence of* Miracinonyx

complications—but at the same time the ability to perform one task is then often co-opted into the needs of another.

A leopard climbing up a more or less vertical tree trunk (figure 4.39) proceeds with a series of bounds in which both forelimbs and both hind limbs act together. Lateral movements of the forelimb are important, permitting the trunk to be grasped, while the hind limbs continue to move in line with the body. The muscles controlling flexion and extension of the spine are also very important during climbing. Compare this posture with that of *Megantereon* pulling down a horse, as depicted in figure 4.40.

FIGURE 4.39 *Principal muscles used by a leopard in climbing*
This figure should be compared with the following one, because the important muscles that allow a leopard to climb a tree are similar to those that permit cats to grasp prey.

Walking

It would seem self-evident that similar-looking animals will move in similar ways, whether leaping, walking, or running, and such similarities to modern forms are of course the basis for the reconstruction of movement in fossil species. Occasionally, however, we can supplement our inferences about likely movement based on skeletal reconstruction with some more direct evidence. It is possible to tell a great deal about an animal from a careful examination of its footprints in combination with a knowledge of its skeletal appearance.

Footprints of fossil cats are unfortunately rather rare, and are usually confined to those of living species from the not-too-distant past. However, some recent discoveries of Miocene age at Salinas de Añana in the province of Alava in northern Spain appear to be those of five individuals of *Pseudaelurus* moving slowly and casually. The tracks (figures 4.41 and 4.42) show an animal about the size of a large European wildcat, with a digitigrade stance. The central pads of the front and rear paws are longer in proportion to the overall foot size than is normal in cats today, implying a larger foot-surface contact with the ground, but the overall effect is highly compatible with the living pattern of locomotion and parallels the modern appearance of the skeleton of small species of *Pseudaelurus*.

One interesting detail about the Spanish tracks is the fact that four of them are in parallel, implying that four animals traveled together. Whether this represented four unrelated individuals, a family with two juveniles, or perhaps a mother with three large cubs is unclear, but any of these possibilities points to social activity.

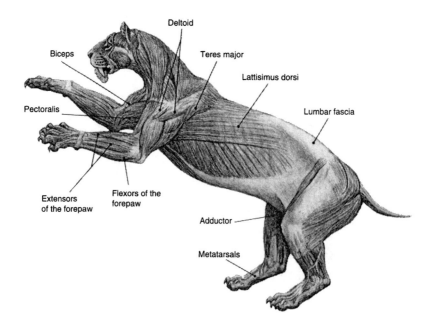

Deltoid

Biceps

Teres major

Lattisimus dorsi

Pectoralis

Lumbar fascia

Extensors
of the forepaw

Flexors of the
forepaw

Adductor

Metatarsals

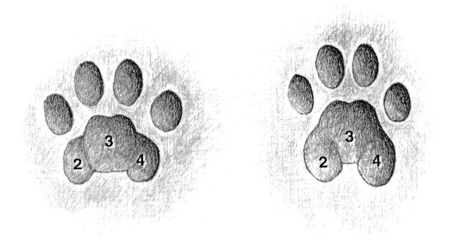

FIGURE 4.41 *Comparison of pad pattern in the Salinas de Añana footprints and a modern cat*
The pad pattern of the cat that left the Salinas footprints (*right*) shows well-developed interdigital pads 2 and 4, which project relatively far backward. In modern cats (*left*), interdigital pads 2 and 4 are reduced and pressed to the sides of pad 3, which produces an overall shortening of the composite pad. A reduced interdigital pad surface is a general trait of the more cursorial carnivores.

FIGURE 4.40 Megantereon cultridens *pulling down a horse*
The killing technique of the dirk-toothed cats very probably involved bringing prey down to the ground before biting at it, unlike the living pantherines which often bite the standing and struggling prey animal in the throat, the muzzle, or even the rump. Such behavior would have been too risky for the long canines of *Megantereon*, and to subdue a large ungulate (a small horse like this would be among the largest prey for this cat) it would have had to exert an enormous muscular force.

Among the most important muscles involved in such activity were the flexors and extensors of the forepaw, which form an important mass of the animal's forearm; the muscles that adduct, or pull in, the arm, such as the pectoralis; and those that abduct, or pull out, the arm, such as the deltoid. There is evidence from the insertion areas on the bones that all these muscles were extensively developed in smilodontine cats such as *Megantereon*.

A further important movement, the flexion of the arm, is carried out by the biceps muscle. This muscle was evidently well developed in *Megantereon*, and its action was aided by the combined effects of two others, the teres major and the latissimus dorsi. The first arises from the posterior surface of the scapula, or shoulder blade, and the second from the backbone and lumbar fascia. They join together and insert on the inside of the humerus. Since the back of the smilodontines is shorter than in feline cats, the contraction of the latissimus dorsi is more effective, as is the pull of the large muscle masses of the back lying under the lumbar fascia (shown in the sequential reconstruction of *Homotherium*, figure 3.11).

In the hind limb the powerful adductor muscles and the shortened metatarsals improve the stability of the cat when it is wrestling with its prey.

FIGURE 4.42 *The Salinas de Añana carnivore tracks*

This drawing shows a reconstruction of one of the cats that left the Salinas footprints. The trackways correspond to animals that were either walking or, as shown here, trotting at a moderate speed (less than two meters per second). Many of the felid trackways show the animals moving in diagonal sequence gaits, which means that each hind footfall was followed by a fore footfall of the opposite side. This is the way in which modern cats move when using a trot or a walking trot. The various measurements of the trackways show that the Salinas cats were proportioned much like some forest-dwelling species of living cats, with long backs and relatively large paws. The gait was kept unchanged for many meters, and one can almost see the small animal moving purposefully along the ancient lake shore much as a modern cat would do.

CHAPTER 5
Behavior and Ecology: Hunting and Social Activity

CATS ARE FLESH EATERS, AND, AS WE HAVE SEEN, MANY OF THEIR physical features show adaptations toward obtaining flesh by capturing and killing animals that they then consume. But obtaining food requires behavioral adaptations as well as purely physical abilities, and in many cases these appropriate patterns of behavior have to be learned. Mothers of young cubs appear to train their offspring by capturing live prey for them and then supervising activities as the cubs attempt to catch and kill the animal. Domestic cats, even when adult, may be seen to capture and then play with mice, shrews, birds, and small rabbits. Such events are often considered by human observers to be an example of deliberate animal cruelty, but in reality they are no more than a continuation of early training behavior.

Killing, while doubtless important for long-term success, is not the only way to obtain meat, and we should not naively assume that all cats in the wild operate as noble hunters. Most carnivores are opportunistic feeders, and they will take a meal wherever and whenever they can get it. In many cases, such opportunities come from driving off other species that may have made the actual kill—a situation commonly observed in some parts of Africa, where lions may scavenge from the kills of hyenas, leopards, and cheetahs.

In this chapter we therefore examine some of the social and behavioral factors that go together to make the larger cats effective hunters and scavengers, and we consider the implications for our understanding of the activities of fossil species.

SOCIAL GROUPINGS

Most of the living cats lead largely solitary lives, coming together for mating and little else. The offspring live with their mothers for some time until they are equipped to fend for themselves (which may take up to two years), but such small family groupings are the only ones to be seen in the case of most species for most of the time. The major exception is the African lion, which may exist in prides of up to a dozen adults with offspring (figure 5.1); but even here, the structure of the pride is by no means a random grouping of individuals.

The overall level of social activity is reflected in the means by which food is obtained, and there seem to be strong interactive links between the two. Among the carnivores in general, the species that hunt in packs are those with more complex and intense patterns of social behavior—a situation demonstrated well by the wolf, the African hunting dog, and the spotted hyena, for instance. In the case of the spotted hyena, the clan in suitable circumstances may contain as many as 80 individuals. It therefore comes as no surprise, given their solitary social activity, to find that most living cats are solitary hunters, and that the major exception is once again the African lion. It is worth looking at the structure of African lion society in a little more detail.

The basis of the pride lies with the females, often related to each other, who do most of the hunting and take great care of the young on a broadly communal basis. As we have already seen, like most cats, and many mammals, lions are sexually dimorphic and the males are considerably larger than the females. The adult males of the pride, numbering perhaps two or

FIGURE 5.1 *A pride of lions relax in the shade*
Lions may often be seen relaxing in this way, perhaps having just fed or perhaps simply conserving their strength and keeping cool during the heat of the day. Much of the social activity of the pride takes place at such times, with cubs playing and feeding and adults either sleeping or grooming.

three and often related to each other (but not necessarily to the females), will have taken over the pride from previous males whom they will have killed or driven off. Cubs fathered by the previous males are likely to have been killed, and any newborns seen with the females will have been fathered by the incumbent males. As male cubs mature they tend to be driven out, but maturing females may stay and join the pride more permanently. The adult males will defend the pride against takeover by others, who may roam in all-male groups in search of prides that they can take over. Adult lions therefore exist either in prides of mixed sex or in all-male groups; only in old age are they likely to live in isolation, and the life expectancy in such circumstances is extremely short.

This intensive social behavior is in turn reflected in cooperative hunting techniques whereby a large antelope or a zebra may be cut out from the herd and attacked by several adult females at once (figure 5.2). The males tend to take little part in such activities; for one thing, they are larger than the females, and they find it more difficult to conceal themselves. However, they are quick to assert their position when it comes to feeding at the kill, and to use their size and strength to ensure their share. From their perspective such behavior makes perfectly good sense: why risk injury or suffer exertion when food will be made available by others? However, in their all-male groups they are frequently forced to feed themselves, and here they may act cooperatively to kill even larger animals such as male buffalo (figure 5.3). Indeed, in some areas of Africa the males appear to specialize in such cooperative hunting of these aggressive and dangerous prey, and in so doing they underline the flexibility of the behavioral repertoire. This flexibility may even be employed at times to the advantage of the

FIGURE 5.2 *A group of pride females attack a zebra*
Although any one of these lions could capture and kill a zebra, joint actions such as this are common and more efficient. While a zebra may shrug off a tackle from a single lion, a situation like the one depicted can really have only one outcome.

FIGURE 5.3 *Two lion males tackle a buffalo*
Like the females, pride males or members of an all-male group may act together
on occasion. However, when such large and aggressive prey are attacked the outcome is less certain, and injury is perhaps more common.

pride, when the combined action of the males provides the occasional
larger item or perhaps tips the balance where the efforts of the females
alone are not enough.

Of course the actual size of the prey taken depends on local conditions
and availability. In Lake Manyara National Park in Tanzania, for example,
George Schaller found that buffalo, *Syncerus caffer*, were the most abundant
large prey species and made up 62% of lion kills. Amazingly, 81% of these
buffalo were adult males—simply because once separated from the herd, as
such older animals often are, they make an attractive target. In contrast,
lions of the Kalahari have to survive on generally smaller prey because large
antelope are relatively scarce.

A taste for large bovids is not exclusive to modern lions, for there is evidence that American lions of the Pleistocene included them in their diet.
A remarkably complete, naturally freeze-dried mummy of a bison found in
Alaska and shown here in figure 5.4 bears clear marks of claws and teeth. A
study of these marks by Dale Guthrie shows that they were not caused by
Homotherium, the other large felid present in the area, because its teeth
would have produced a ripped rather than a punctured wound. Other evidence suggests that the kill was made in the autumn, and that glacial temperatures caused the body to freeze before it could be consumed. This

FIGURE 5.4 *Blue Babe*
The frozen remains of a Pleistocene bison, nicknamed "Blue Babe" by its discoverers, show the clear marks of having been killed by lions, with puncture wounds in the hindquarters as well as the muzzle. The actual kill may have been achieved by the typical lion bite shown in figure 5.16.

would suggest that the kill was not made by a large pride. However, at least one attempt was made to eat from the frozen carcass, for the lion that did so broke one of its carnassial teeth in the process and left the fragment in the skin.

The sheer size of the individual lions, and the weight of numbers that can be amassed by a pride, means that a second option is open when it comes to obtaining food (figure 5.5). Spotted hyenas are themselves intensely social hunters, able to chase and bring down zebras and larger antelopes when operating in hunting groups of up to 25 individuals. The prey of spotted hyenas may therefore offer an attractive source of food to any other predator, if it can be reached before the hyenas have had time to consume it. Spotted hyenas will guard a kill with considerable aggression, but even a large number of them can do little against a pride of adult lions, and in parts of eastern Africa they have been observed to lose up to 70% of their kills to lions on a regular basis.

African lions live today mainly in more open areas, in conditions ranging from savanna woodland to semiarid, and their social behavior makes good sense in such an environment. Cooperation in hunting and rearing the young increases efficiency at all levels, and ensures that larger prey can be obtained than would be possible for an individual to tackle (figure 5.6). Of course, in times of food shortage, group living may also mean that there is not enough food to go round, and in such cases it is often the cubs that suffer since they cannot compete with the other pride members. At such times the males have been observed to give preference to the cubs at the kill, driving off the females to ensure access for the young.

FIGURE 5.5 *An adult lion confronting a spotted hyena*
Lions exhibit a low tolerance of hyenas at kill sites, and will often give deliberate chase.

In contrast to the lion, the savanna-dwelling cheetah often lives and operates alone. The specialized hunting behavior of this species renders cooperation difficult. However, the situation is confused by reports of several cheetahs at times acting in concert; in the Serengeti, for instance, there are reports that 60% of adult males may live in permanent coalitions of 2 to 4 individuals, while only 40% lead more solitary lives. There is evidence that such group living may be beneficial: males in the group are reported as averaging 10 kg heavier than solitary animals. In addition, it seems that family groups of subadult offspring and mother may often hunt together, perhaps giving the impression that associations of adult cheetahs are more usual than is actually the case. There is also evidence that siblings may stay together for some time after leaving the care of the mother, and such partnerships confuse the picture further. The initial training period of the cheetah is extremely important to the animal because the hunting method is so specialized, and it is possible that strong bonds are formed at that time. It was a common practice in former times in Asia for cheetahs to be used in hunting, but it was recognized that young animals taken too early from the wild would not have learned to hunt properly—such is the importance of the instruction received from the mother.

Leopards, tigers, and jaguars tend to operate in more closed terrain, where they can stalk and close in on their prey more easily and where greater numbers would be of little value at the kill. Social interaction and cooperation in hunting are therefore usually minimal in these species, and the marking of territory (figure 5.7) is a significant feature of their behav-

FIGURE 5.6 *A group of lions feeding at a carcass*
An adult giraffe is a formidable opponent, rarely attacked by even the most hungry lion, but juveniles are taken. Once downed, such an animal will provide a meal for several lions.

FIGURE 5.7 Nimravides pedionomus *spraying urine on a tree trunk*
The early species of the American genus *Nimravides,* found in Clarendonian deposits of Miocene age, are somewhat smaller and more primitive than the later forms such as the lion-sized *N. catacopis* from Hemphilian deposits. *N. pedionomus* was about the size of the smaller, southern subspecies of modern tiger and was built much like those animals. Like many living cats it was probably territorial, and we have depicted it here in the act of marking the boundaries.

ior. This solitary existence is also true of the puma, and although it lives in a variety of habitats it seems that all offer little stimulus toward cooperative behavior. The same may be said for the snow leopard, in which a combination of open terrain and highly dispersed food resources make individual hunting the only serious option. But is the more solitary lifestyle of these animals real, or only apparent?

The tiger, as we mentioned earlier, appears to exhibit social tendencies, and snow leopards in captivity seem quite capable of living together amicably. It would seem that sociality falls within the range of possible behavior patterns for both species, and that cooperative arrangements could become the norm in different circumstances. Each male among leopards, jaguars, and tigers has a large territory that tends to overlap with the smaller territories of several females (plate 11), and observations of tigers and leopards suggest that the territorial male may be more tolerant of the females and the cubs that it has fathered than was previously thought. Family groupings have even been seen, and in some cases leopards have been observed to mate while the female still has cubs. There is also now some evidence that, as with lions, it is the juvenile leopard males that are likely to be less tolerated by the mother and forced to disperse, while the female cubs may remain in the vicinity for longer and even establish territories adjacent to hers.

Both solitary and social cats are markedly territorial, and mostly sedentary. They share a repertoire of signals with which to mark their territory, although there is some variation depending on species and habitat. Olfactory, vocal, and visual messages are used to warn trespassers, and the importance of smell seems to be greater for communication with other species members than it is for hunting. Urine spraying, as depicted in figure 5.7, is the most conspicuous behavior in olfactory marking, but members of prides or family groups also rub heads and necks in greeting, thus exchanging smells from facial glands and acquiring a common, familiar scent.

Loud and far-reaching vocalizations, such as roaring, are mainly used for territorial advertising, but the range of vocal expressions is very wide in cats. Some sounds, like the purring of the kittens of smaller species, actually seem intended to be heard at only very close range, within the intimacy and security of the family group.

The most obvious visual territorial signal is the exhibition of the animal's own body, with its range of markings such as manes, beards, spots, and stripes acting as a clear recognition sign. Facial markings aid in expressing friendly or hostile intentions, while the development of manes and beards, coupled with the sheer size of the territorial male, offer clear discouragement to intruders. More subtle visual signs include the droppings or scats, when not covered, and the claw marks left on the trunks of trees. These are seemingly recognized by other cats, although some authors have expressed doubts about whether such marking is intentional. But there are scent glands in the paws of a cat, which would ensure that any claw marks are not only seen but also smelled.

The variation in social behavior within and between the cat species serves to underline the fact that the Carnivora as a whole are a behaviorally adaptable order, able to adjust their patterns of activity to a marked degree to suit circumstances. Wolves, spotted hyenas, and lions, all at the extreme end of the range for communal-living species, adapt their group size—and the resultant patterns of social interaction and hunting behavior—in line with local conditions, including vegetational cover and prey availability. This range of behavior is exemplified by the Asian lions of the Gir reserve, where denser vegetation than that usually encountered by African lions reduces the need for a tight pride structure: for much of the year these males and females tend to live in separate groups, meeting only for mating. Therefore no single arrangement typifies the social behavior of such animals. A further aspect of felid sociality is that male and female groupings seemingly stem from different causes. In lions we observe both male coalitions and female groups, but in cheetahs only male coalitions are evident. The naturalist Thomas Caro has suggested that the main factor driving female associations in large cats may be the presence of high densities of prey weighing between one and two times the weight of the adult female, while male coalitions may be driven by a high density of females concentrated in well-delimited locations.

But the range of behavioral possibilities is not infinite for any given species. Why the marked differences between the larger carnivores? There have been suggestions, most notably by the German zoologist Helmut Hemmer, that the level of social complexity is correlated with relative brain-body size in carnivores and therefore, by implication, with learning capacity. Among the living cats, the lion and the tiger both have the greatest relative brain size, followed by leopards and jaguars and then pumas, which has been taken to equate with what we know of their social interaction. But what do we make of such arguments if the degree of (or apparent potential for) such activities has been, as now seems to be the case, underestimated? All the observed patterns taken together begin to look rather like primitive, perhaps even incipient, pride behavior, and suggest that ecological factors have served as the triggers for more intensive development in some species while causing little impetus to change in others. The brains of cheetahs are big enough to enable them to learn a complex hunting technique from their mothers. The fact that some cheetahs may live together suggests that there is no major, innate impediment to doing so. But the cheetah has gone down the path of hunting by high-speed chase, necessitating a particular capture technique and limiting the size of prey that can be taken. There would be no point in living in a larger social group in this situation: greater numbers using the same hunting technique could still not bring down larger prey, and more mouths to feed at a kill would mean less to eat. Brain size may therefore not be a particularly good predictor of behavior across the family as a whole.

Arguments about the limitations of sociality in living cats lead us quite naturally to the question of possible social behavior in fossil cats. Here it is easy to be seduced by the attractive notion that relative brain size can tell us something, since it provides a measure for comparison between fossil taxa, with living species as a baseline. However, the paleontologist must proceed with even more caution. It is reasonably easy to estimate brain size by measuring the capacity of the empty skull, although that itself is not the actual figure aimed at because the inside of the skull is not filled by the brain tissue. Yet even if we could estimate the actual size of the brain, it is really the amount of cerebral cortex, the outer layer of the brain where the higher functions are located, that would be of interest to us, and that is impossible to infer directly. And in any case, even if we disregard these sources of error, the question concerns *relative* brain size and thus it is equally important to estimate the body size. This is much less easily achieved, because it involves measurements of bones and then various calculations based on assumptions about bodily proportions that may be less accurate than we assume. Sexual dimorphism is likely to be a confounding factor in such exercises, since the males of many of the cat species are considerably larger than the females (figure 5.8). Adult tiger males may be more than 1.5 times bigger than females of the same population, to say nothing of possible clinal variation across the range of the species. If the correlation between brain size and sociality is itself in question, the whole exercise is fraught with difficulty. Perhaps the best that we can say is that a level of relative brain size in a fossil cat similar to that of an equivalent living species implies a broadly similar range of possible social complexity.

The whole issue can perhaps be seen in context with some examples. American fossil lions, although now usually considered to belong to the same species as the extant lion, actually show a greater relative brain size than their living relatives when these calculations are done. Should that mean that they were even more social, or perhaps capable of even more complex behavior? In contrast, the large saber-toothed species *Smilodon fatalis*, abundantly represented in the Rancho La Brea deposits, has a relative brain size similar to that of the living leopard and jaguar. This, together with the evidence of a high incidence of lesions on the bones of the *Smilodon* specimens from the same localities, has been used to argue for a low level of social interaction and a fairly high level of intraspecific aggression in the species. However, other lines of evidence suggest that this may be something of an oversimplification. Maternal care is likely to have been important for the young while the dentition was developing (figure 5.9). Further, many of the injured individuals show evidence of healing, meaning that they lived for some considerable time after the injury, while many others seem to have been of advanced age or suffering from a number of pathological conditions that would have made hunting very difficult; that such animals survived is by itself a strong indication of social behavior, since we know from studies of lion prides that incapacitated individuals may be well tolerated and given access to food.

FIGURE 5.8 *Couple of* Machairodus giganteus *mating*
It is very likely that, as in all living felids, mating was a very noisy activity for machairodontines, which may have been roaring cats. In this drawing we can also appreciate the size difference between the sexes, a marked feature of *Machairodus*.

FIGURE 5.9 *Family scene of* Smilodon *female with young*

All living cats give birth to young that are heavily dependent on their mothers in the initial stages of life, and fossil remains of extinct species show similarly undeveloped cubs that would have required close attention. The family bonding that results from such behavior may underpin future social behavior patterns between adults.

A further argument against the solitary and aggressive lifestyle of *Smilodon* has been put forward based on the sheer numbers of individuals recovered at Rancho La Brea and a few simplifying assumptions. First, it seems likely that the cats became mired in attempts to reach already-trapped animals or their carcasses. If the number of large herbivores found gives an approximation to the maximum number of trapping incidents (a plausible suggestion), then several individuals of *Smilodon*, perhaps ten or so, were also trapped during their efforts to obtain each of those herbivores. Since it is unlikely that every *Smilodon* attempting to reach a trapped herbivore did indeed suffer the same fate, the implied number of animals in the vicinity at any one time, even if the trapping incidents were themselves infrequent, is really too high to support an argument for a solitary lifestyle, since such a lifestyle would imply discrete territories and would reduce the number of animals able to congregate in the area.

So far as other taxa are concerned, we suggest some possible aspects of behavior in *Machairodus* in figures 5.10 and 5.11. The likelihood of social interaction by members of the genus *Homotherium* has also been put forward by several authorities. It is certainly true that brain size alone would not argue against such behavior, while the evidence that it hunted large prey even though it was less equipped to subdue its quarry than *Smilodon* or some of the pantherine cats would point to the possibility of group activity. The cursorial traits in its skeleton may point to life in rather more open environments, and it would make ecological sense for it to have operated in some kind of pride in such conditions. The denning behavior that seems to have occurred at Friesenhahn Cave has been seen as evidence of solitary behavior based on living species—but then European lions have been found in some numbers in caves in France, Poland, England, Germany, and elsewhere in circumstances that point to occupation by more than just single animals.

But there is another side to this. At least part of the difficulty in reconciling the differing interpretations of locomotion in *Homotherium* has been the need to incorporate into the discussion the evidence for what appears to have been predation on juvenile mammoths from Friesenhahn Cave (figure 5.12). The evidence for the juveniles is of course the presence of more than 70 young mammoths in the deposits. But what killed them, and how did they actually get there? In the absence of a living representative, kills of elephants by lions seem to offer our best parallel, and the observations of Derek Joubert in Chobe National Park in Botswana are one of the best sources. He found that elephant meat made up as much as 20% of the diet of one pride during 1990, and although much of that meat was scavenged from natural deaths, some did come from kills. Here the target was young animals between two and four years of age—animals whose curiosity is likely to take them some way from the herd, but not babies, which are closely guarded by their mothers. This finds parallels among the Friesenhahn sample, as noted by Rawn-Schatzinger, where

FIGURE 5.10 Machairodus giganteus *couple and* Tragoportax amalthea
The question of sociality and the extent of cooperative action underlies the reconstruction of lifestyles for all extinct cats. The array of prey animals available to
Machairodus giganteus in Eurasia included heavy beasts larger than buffalo, such as
the sivatherine giraffids, as well as medium-sized, fast-running antelopes like
Tragoportax. Hunting in groups would have been advantageous for the cats in both
cases. While a greater "firepower" would have allowed the tackling of very large
prey, the smaller and more fleet-footed antelopes could have been "trapped" by
cooperative tactics that drive the prey into the claws of a pride member waiting in
ambush. Such cooperative hunting is usually associated with lions, but it has also
been observed among lynxes and tigers.

most of the juveniles were around two years old. Such kills usually involve
several lions and can take up to one and one-half hours, since the lions
cannot apply an effective throat bite or muzzle clamp. Here *Homotherium*
may have had a real advantage if it was able to slash through the thick skin
and induce massive blood loss.

  The other major predator at the site was the dire wolf, *Canis dirus*. This
animal was the size of a large gray wolf, but much more powerfully built,
with a large head and enlarged teeth. However, it may well have been
unable to immobilize a young elephant and prevent it from rejoining the
herd, unless it acted in particularly large packs. On balance, *Homotherium*
may be the more likely candidate as the predator if the mammoths represent kills rather than natural deaths (and natural deaths seem rather
unlikely as an explanation for such a high proportion of juveniles).

FIGURE 5.11 Machairodus *couple with* Birgerbohlinia schaubi
The giraffid *Birgerbohlinia schaubi* was a large animal, and if the adults were taken by *Machairodus* then a combined effort would have been essential.

But how did the remains get into the cave? Even a juvenile mammoth was no lightweight, and the presence of teeth implies that skulls and/or mandibles were brought in. Lions and leopards drag carcasses of considerable size into places of safety, and leopards of course even take them up trees, but elephants are not among the examples quoted. Moreover, the dentition of *Homotherium*, despite the forward setting of the incisors discussed in the last chapter, does not look best suited to grasping and dragging such very heavy carcasses or parts of carcasses. It is perhaps at least as likely that the mammoth remains at Friesenhahn Cave were brought in by the powerful dire wolf, and if that were the case then the mammoth remains might shed no light at all on the carcass-carrying and locomotory behavior of *Homotherium*. Against that theory, however, is the likelihood that if the mammoths represent scavenging by the dire wolf then one would expect to find other prey species as well, giving a wider picture of the local fauna and the dietary range of the dire wolf. In short, such a bias toward a single species is rather implausible for a scavenger, and we seem to be left with little alternative to accumulation by *Homotherium*—although the question of dismemberment and actual transportation still warrants some attention, since it is hard to believe that whole carcasses were dragged into the cave. What does seem clear is that if *Homotherium* was the predator, then solitary activity is extremely unlikely.[*]

[*] Since completion of this manuscript, Curtis Marean and Celeste Ehrhardt have published a detailed discussion of the patterns of body-part representation and damage to be seen on the Friesenhahn mammoth bones. They conclude that there is considerable evidence in these patterns for disarticulation of young mammoths by *Homotherium*, followed by transportation and consumption in the cave.

FIGURE 5.12 Homotherium serum *and young mammoth*
Although a juvenile mammoth would have been a suitable prey for *Homotherium*, such young animals are rarely found far from the rest of the herd in living elephants, and it is likely that mammoth herd structure would have been every bit as tight. There would be little advantage in making such a kill only to be chased off by an irate matriarch. However, young mammoths between two and four years of age would probably have been less closely guarded, and their curiosity may have taken them dangerously far from the herd as it does with modern elephants.

THE KILL

Unlike other large predators such as the wolf or the spotted hyena, which essentially kill their prey by pulling it down and eating it, cats tend to dispatch their chosen victim first, and only then do they begin the process of consumption (figures 5.13 and 5.14). We can see in the drawings how the comparative anatomy of the head and forelimbs of the hyena and the lion reveal these differences. The large claws of the cats, and the powerful forelimbs, are ideally suited to catching their prey and retaining a hold while wrestling it into a position where the canines may be employed in a bite. Dogs, like hyenas, lack large and retractable claws and their canine teeth are less developed, so that sheer weight of numbers is often the most important factor in bringing down an animal.

The precise method of capture and dispatch depends on the size of the cat and the size of the prey, and it is here that we see the most obvious interaction of innate and learned patterns of behavior. Domestic cats, as anyone who has ever kept a kitten will know, show an early interest in chasing anything that moves, and will go through an elaborate sequence of crouching, wriggling the hind quarters, and then pouncing in an impressively organized manner. What they seem less sure of is what to do with the object once they have seized it. The neck region, if there is one, appears to be sought instinctively, but the best method of capture of live and moving prey, and the specific orientation of the bite, seem to benefit from guidance. We referred earlier to the habit of female cats of bringing live prey for the young to practice capture and killing, and it is through this that the basic instinct for capture appears to be directed and developed.

The extent of the basic killing instinct may enable us to place the frequent reports of excessive killing by predators into some perspective. Most of us have heard tales of the extent of damage that may be wreaked by a fox in a chicken pen, far beyond the possibility of consumption, and we may even have sympathized with the interpretation of wanton killing and the calls for eradication that tend to follow such incidents. Spotted

FIGURE 5.13 *How dogs and hyenas hunt and capture prey*
These two figures illustrate how the ways in which dogs and hyenas tend to capture and subdue their prey differ from the methods used by cats. Dogs and hyenas have only moderately large canine teeth and short, nonretractable claws in relatively immobile paws. They do not reach as great a size as the largest cats, and yet they may hunt very large prey—as in the case of wolves that tackle moose of the genus *Alces*. Success is usually attributable to a combination of their stamina, to maintain a long chase, and their technique of bringing down the exhausted prey by sheer weight of numbers, as illustrated by the hyena clan attacking a zebra. In most cases, dogs and hyenas kill the prey by eating it.

FIGURE 5.14 *How cats hunt and capture prey*
In contrast to the dogs and hyenas, cats usually hunt singly (although lions hunt in prides, and some other large cats may act in family groups). The prey is seized after a stalk and a relatively short dash—except in the case of the cheetah, which has its own special method. Small cats may deal with small prey by means of a neck bite. Larger cats can tackle very large animals, and a wrestling match may follow contact. The long claws permit a firm grip to be established, while the long canines may be used to bite into the throat, or the animal may simply be suffocated. The prey is therefore usually dead before feeding commences.

hyenas have been observed to act in a similar manner when faced with a surplus of gazelle calves, and pumas have long been known to slaughter domestic animals in a seemingly random and senseless manner. What is clearly happening, however, is that the killing instinct in each case is being provoked, without the intervention of a full stomach to "switch off" the response. Lions and cheetahs have been observed to cut from eating a carcass to chasing another animal if it happens to have blundered upon them feeding and to have provoked the hunting response, particularly if it runs.

For smaller prey, a bite at the rear of the neck will suffice, driving the long upper canines between the vertebrae and severing the spinal cord. Jaguars (figure 5.15) have been observed to employ a somewhat more sophisticated method on suitably sized prey, the capybara, taking the head in the mouth and feeling for the ears with the upper canines before biting down and into the brain.

Larger animals, especially the larger ungulates and most particularly those with horns or antlers, require different approaches. Among the favored techniques of the lion, for example, are biting the throat, and simply clamping the mouth over the whole muzzle of the animal, as shown here in figure 5.16. These may be achieved from a position where the cat is effectively hanging from the neck and shoulders of its victim, so that the killing bite is in place while the animal is still standing. In both cases death results from suffocation rather than the violent and bloody end that is often assumed. Of course if the other members of the lion pride have been involved in the kill it is likely that up to half-a-dozen other cats will be swarming over the body by the time that it is actually brought down, so that various parts of the body may be attacked at once. But even then, one lion is likely to have selected the head or neck as the preferred point of attachment, and it may even be joined by one of its pride companions in that position once the ungulate is on the ground. Descriptions of kills suggest that prey may be more or less in a state of shock by the time they are brought down, and although some animals may put up a fight it seldom seems to last for long. Instead, they are often reduced, as has been said, to being witnesses to their own execution.

The equally large tiger operates in much the same manner as the lion when it comes to individual kills, and females with subadult young may

FIGURE 5.15 *A jaguar killing a capybara*
As the drawing shows, the jaguar often bites the skull of the capybara, reaching into the brain through the ears. This technique depends for its success upon the precise orientation of the canines.

replicate some of the features of the communal pride behavior at the kill. In some cases the tiger has been observed to twist the neck of the prey, and it has actually been heard to break it. Smaller cats, such as the leopard, jaguar, and puma, use essentially the same techniques of prey dispatch on animals around their own body size, although most get closer to their chosen victim before the final dash and capture than does the lion.

The stalk undertaken by some of these more solitary cats brings us back to the question of intelligence and the potential for socialization raised by the debate about brain size. The hunting technique of the leopard often requires a degree of intelligence and apparent deliberation no less than that needed for cooperative hunting. The leopard does not merely wait for something to blunder past, or even approach the intended prey in a rapid and straightforward manner. Instead, it often circles entire herds of ungulates, seeming to anticipate their movements and placing itself in the likely path. These complex excursions may take hours, with the leopard sometimes abandoning the first strategy and starting afresh from the opposite direction.

The sophistication and learned aspect of some solitary techniques for prey capture are exemplified by the tigers of Ranthambore in India (figure 5.17), which became famous for their spectacular habit of rushing into the water to catch the sambar deer off guard while they fed on aquatic plants. But this technique was unknown before the early 1980s, when an athletic male nicknamed Genghis seemingly invented it. No mention of such behavior occurs in the literature on tigers collected over the past two hundred years. Even after Genghis died the technique lived on, learned by others and itself evidence for some degree of social interaction. Sadly, this tradition may soon be lost, since intensive poaching in the region has nearly wiped out the tigers of Ranthambore.

FIGURE 5.16 *A lion dispatching an antelope with a muzzle clamp*
Such a method of killing is relatively bloodless and can be extremely effective.

The major departure in capture technique is seen in the case of the cheetah (figure 5.18). Like many of the other cats it is perfectly adept at the stalk to bring itself closer to its eventual prey, freezing into absolute stillness if its movements seem about to be detected. But the final rush takes the form of a high-speed chase, often over several hundred meters, during which the twists and turns of the prey (usually a small antelope) are relentlessly followed. The chase may begin with a gentle trot from as far as 250 meters away, so that by the time the cheetah is seen it has already gained momentum. Even so, the acceleration, once the cheetah commits itself to the selected victim, is astonishing, as though the cat is able to change gears.

The capture is achieved at high speed, usually not by a leap onto the back of the animal but by clawing at one side of the rear of the prey and pulling backward in a complex and carefully coordinated maneuver. This causes the prey to lose balance and collapse, and usually results in its tumbling over. The large dew claw on the inside of the cheetah's front paw is employed in this technique, in effect "hooking" the back leg of the unfortunate antelope. Precise observation in the field is difficult, owing to the clouds of dust within which the final capture usually takes place, but slow-motion films of cheetahs in action, backed up by studies of captive animals, show that the cheetah actually operates most effectively at speed. At a slower pace it seems unable to cause its prey to stumble and lose balance to a sufficient extent, perhaps because the pursued animal is in greater contact with the ground when not leaping at high speed. Experienced adult

FIGURE 5.17 *Ranthambore tiger*
This illustration shows the spectacular hunting technique developed by the tigers of Ranthambore National Park in India: the cat rushes into the water, and the sambar deer feeding on aquatic plants are caught by surprise.

FIGURE 5.18 *Cheetah hunting sequence*
Like other cats, the cheetah uses a cautious stalk to approach its intended prey
(*first drawing*), but instead of the typically low, crouching posture of the leopard or
the lion it simply lowers its head and flexes its legs slightly. Contact with the prey
usually occurs at very high speed, and the cat throws the antelope off balance (*second drawing*) by "hooking" a hind limb with the internal dew claw of its front paw.

cheetahs have been observed to capture some 70% of the prey chased, a figure almost double that for the efficiency of lion predation.

As an alternative, the cheetah may grasp at the flanks of the fleeing prey with both front paws, dragging the animal down rather than tripping it (although a tumbling effect also tends to occur, since the seizure may still come during a fairly high-speed chase). If there is any degree of cooperative activity between family members, it is at this point that it becomes important, helping to subdue the prey before it can regain its feet and attempt to run again.

Such a hunting technique is possible only in suitable terrain, where open grassland to permit the necessary sprint is interspersed with enough vegetation to provide cover. In addition, the size of the prey is a crucial factor. A lion, or even a leopard, can kill an animal greater than its own body size, especially in a group hunt, although a size at or about that of the predator seems to be a general maximum among cats. But the cheetah is often near exhaustion at the end of a successful hunt—especially if it has involved a final chase of 400–500 meters, its apparent limit for an all-out sprint. It is thus vital to the cheetah that the prey be small enough to subdue easily, usually by the common technique of strangulation. Small gazelles are therefore favored, with a maximum body weight of up to 50 kg; the favorite prey, Thomson's gazelle, weighs less than 30 kg. At this size they can be wrestled to the ground even if they have started to rise again after the initial capture. Once the gazelle is on the ground, the preferred means of killing seems to be to lie across its shoulders from the rear, to seize the throat, and to twist the neck around so that the prey is almost facing backward over its own shoulder. In this way the horns are directed away from the cheetah, while the closure of the windpipe imposed by the bite is perhaps exacerbated by the twisting of the neck. The gazelle succumbs after a few minutes, usually with very little struggle, and feeding can then begin almost at once, before the almost inevitable appearance of hyenas or other large cats (see next section).

So far as the hunting behavior of fossil cats is concerned, we can probably assume that the same general principles would have operated. Capture would be by a stalk and a pounce, usually with the benefit of thicker vegetation, as practiced by the tiger or leopard; or perhaps by a stalk and a relatively short rush in more open terrain, as we tend to see with the lion. Although the animals killed would themselves have differed in some respects from those alive today, many of them were not so very different from those now living as far as the large cats of the day were concerned. What would have been important was the changes in the overall composition of the prey fauna that occurred over time, but we shall discuss those matters more fully in the next chapter.

We can also probably assume that species with modern representatives, or close relatives, would have operated in broadly similar ways. The giant European cheetah was clearly not equipped for a significantly different method of hunting (figure 5.19). The range of prey available to it was considerable, including a number of species of deer and small bovids, such as *Gallogoral*, of just the right size for a hunter perhaps half as large again as the living cheetah. In North America, the similarly adapted members of the genus *Miracinonyx* would have found ready targets among the endemic pronghorns, the Antilocapridae, as illustrated in figure 5.20.

In many cases, cats take prey near to the upper limit of their abilities, and clearly it is more efficient to do so. But if that is not possible then they simply eat what they can get according to local circumstances. The living cat that most resembles some of the smilodonts in its body build is the robust jaguar, which survives today in Belize on a diet of mainly armadillos, each of which weighs less than 10 kg with armor. This may seem an unlikely diet for a large cat, but that is what is available; the same cat's diet in the past would have been very different (figure 5.21). Dietary shifts have been documented in several cat species, and it seems that they quickly react to scarcities in what may have been a preferred species by turning their attention to another, although success may vary (figure 5.22). They are hardly alone in this; spotted hyenas will happily get by on a diet of flamingos if they have to (but then they do have an awesome and deserved reputation for eating virtually anything). We should be aware of that adaptability when thinking about fossil species (figure 5.23).

Another factor influencing the predatory behavior of any cat species is the composition of the carnivore guild. In modern habitats, we have many examples of large carnivores that change their feeding methods according to the presence or absence of more or equally dominant species. In Asia, for example, leopards become more nocturnal and tend to hunt smaller prey in places where tigers are abundant. In Africa, as we show in the next

FIGURE 5.19 Acinonyx pardinensis *attacking* Procamptoceros
The large size of the Villafranchian cheetah raises the question of what it hunted. Most probably its hunting technique was similar to that of the living species, which seems to exclude very large prey, and it seems likely that adult horses and some of the larger deer were beyond its power. A middle-sized antelope like *Procamptoceros*, which was closely related to the living chamois but bigger, may have been among the largest prey for *Acinonyx pardinensis*.

In the illustration we have depicted the moment after the cheetah has "hooked" the antelope at full speed, and the prey is sliding on the ground. To pursue midsized antelope like *Procamptoceros* or the large, ghoral-like *Gallogoral*, the cheetah may have had to venture into hilly terrain, but that is not as strange as it may seem. Living cheetahs in some areas of southern Africa operate in more irregular terrain than we may usually imagine, and the snow leopard, a master of mountain environments, has bodily proportions more similar to the cheetah than to any other large cat.

section, the presence of spotted hyenas forces leopards to take carcasses up into trees more often, a response that also tends to limit the weight of animals that the cat can safely hunt. In modern, relatively intact ecosystems, the coexistence of several species of large predators is apparently achieved by a complex apportionment, as shown by authors such as Udo Pienaar and Gus Mills in the case of the Kruger National Park of South Africa. Five species (lion, spotted hyena, leopard, cheetah, and wild dog) coexist there to varying degrees and with somewhat overlapping diets, but direct competition and its consequences are minimized because the animals tend to use different habitats and to hunt during different times of the day. Where interaction does occur the lion is dominant over all other species, and hyenas steal an important proportion of the kills of cheetahs; but the leopard avoids the worst of the competition by its preference for riverine forest and its habit of caching food in trees.

FIGURE 5.20 Miracinonyx trumani *and a pronghorn antelope,* Antilocapra americana Although the pronghorns of the family Antilocapridae were very diverse until the later Pleistocene only *Antilocapra americana* survives today, and it may be fairly considered as the North American equivalent of the African antelopes. If *Miracinonyx* did indeed evolve independently of its Old World double, the cheetahs of the genus *Acinonyx*, then the need to catch the fleet-footed antilocaprids may have directed the path of its evolution.

FIGURE 5.21 *A jaguar stalking a group of horses*

The scene depicted here is set in the upper Pleistocene of Chile, and shows the jaguar set to attack a group of short-legged horses of the genus *Hippidion*. Horses became totally extinct in the Americas at the end of the Pleistocene, and were only reintroduced by European colonists, but smaller species would have formed a natural target for predation by many of the large cats.

FIGURE 5.22 *A group of Peruvian lions with a glyptodont*

This scene, set in the Pleistocene of Peru, shows a small group of lions puzzled by an individual of the genus *Doedicurus*, a large and heavily armored relative of the armadillos and sloths. Whether the animal could move quickly enough to employ the expanded, macelike feature of the tail is unclear, but its armor alone is likely to have afforded adequate protection for the adult.

FIGURE 5.23 *A puma attacking a pair of* Macrauchenia
Although somewhat camel-like in general appearance, this genus of South American ungulates had a reduced nasal region that suggests they may have had a small trunk. The young of such animals in particular may have made an attractive target for predation by the puma, as depicted here in a scene set in the Pleistocene of Bolivia.

What we can be less certain of is the hunting behavior of completely extinct species with no close relatives, especially the saber-toothed cats. We have already discussed some of the difficulties in interpreting the hunting behavior of *Homotherium* at Friesenhahn Cave. But even here we can be guided by general principles. The dirk-toothed *Megantereon cultridens*, for example, shown here in figure 5.24 in a series of reconstructions with a horse and a deer, is known to have been a strongly built cat, eminently suited to wrestling its prey to the ground, holding it with its large and powerful front claws, and then perhaps dispatching it with a fairly carefully placed slash across the neck, or even a deep bite, as discussed in chapter 4. Rather than suffocating its prey, as modern cats usually do, the dirk-tooth may simply have allowed it to bleed to death, or it may even have taken the opportunity to sink its teeth further into the neck once the animal was weakened through loss of blood and thus less likely to damage the upper canines while struggling. Shock may have been one of the most potent weapons employed, the combined effects of the chase, the capture, and the loss of blood being sufficient to ensure death.

In all probability the weight limits for a cat like *Megantereon* were not far above those of the similarly sized (and similarly built) living jaguar, which can take prey up to the size of a tapir, domestic cow, or domestic horse. The structural differences in its teeth, skull, and vertebral column are probably related more to the detail of how it killed rather than what it killed. Therefore, if we want to gauge what it may have eaten we have to look at suitably sized prey in the sites where it occurs. In the latest Pliocene Spanish locality of La Puebla de Valverde, for example, about 2.0 Ma old, the most abundant species is a gazelle, *Gazella borbonica*. Next in abundance is the zebrine horse *Equus stenonis*, followed by the deer *Croizetoceros ramosus*. Less abundant, but still important, are a second species of small antelope, *Gazellospira torticornis*; a second deer species, *Eucladoceros falconeri*; and a ghoral-like antelope, *Gallogoral meneghini*. Rhinoceros and elephant are present although rare, but are presumably of little relevance to this discussion. So what should we conclude from this faunal list? One point to stress is the very modern appearance of the assemblage. *Megantereon*, the most abundantly represented felid at the site, had to cope with a very "normal" array of prey species, much like those available to any modern large cat, and we see no archaic, slow, or clumsy creatures upon which it could depend.

Evidently, our dirk-tooth had to make a living from catching and eating bovids, equids, and cervids, and although we know nothing about the relative abundance of those in the living prey population it would be logical to assume that it hunted from among the more common taxa. But the most abundant fossil prey species there, *Gazella borbonica*, may have been generally too small and too fast for it, and more suited to the cheetah, which also occurred at the site. The other prey are also likely to have been sought by *Homotherium latidens*, and by the two hyenas found there, *Chasmaporthetes lunensis* and *Pliocrocuta perrieri*. The former was a catlike species, in terms of general build and teeth adapted more to flesh eating, but the latter was a large, bone-crushing hyena undoubtedly adept at both hunting and scavenging. *Megantereon* may therefore have been most interested in the horse and the deer at La Puebla de Valverde, but it was not alone in that interest.

In general, we can reasonably suppose that at sites where the lion-sized *Homotherium* and the jaguar-sized *Megantereon* coexisted the former would have tended to dominate, perhaps pushing the smaller animal to the more forested sections of the habitat. The later appearance of the giant hyena *Pachycrocuta brevirostris* in the earliest Pleistocene would surely have introduced a further element of stress in the lives of both cats, and especially for *Megantereon*. Although the incisor arc of *Megantereon* would have allowed it to drag its prey, it would probably not have been able to carry it into a tree in the way that a leopard can.

Interestingly, in Chinese sites from Pliocene to mid-Pleistocene age we find unusually small specimens of *Homotherium* in association with some of the largest known specimens of *Megantereon*. In the hominid locality of Zoukoudian, for example, a skull of *Homotherium* has a slightly smaller length at its base than a skull of *Megantereon* from the same deposits. Given the relatively lighter build of the homotherines, we can therefore expect that the smilodontine from Zoukoudian was considerably heavier and more powerful, and may even have reversed the normal dominance relationship between the two. Whatever the environmental circumstances that led to this situation, we may be seeing something of the kind of circumstances that led to the evolution of *Smilodon* in the Americas from a *Megantereon* ancestor.

As a final point here it is worth stressing that we should avoid the temptation to treat our interpretation of any one example from the fossil record,

FIGURE 5.24 Megantereon *hunting sequence with horse and deer*
The illustration of the likely hunting technique of *Megantereon* shows the cat stalking a horse in fairly open territory, closing in with a short dash, and then wrestling the animal to the ground at the shoulders (as we saw in figure 4.40). Once the prey was down and held fast it would then have been relatively easy to inflict a damaging wound to the neck of the kind shown in figure 4.24. Such a bite would have been impossibly dangerous for the cat while the horse was upright and struggling. In contrast, the reconstructed attack on the male *Eucladoceros* deer shows just how difficult it would have been for the cat to tackle such an animal because the antlers would have tended to get in the way.

FIGURE 5.24 Megantereon *hunting sequence with horse and deer (continued)*

even if correct, as entirely typical of the behavioral pattern of the species in question. As we hope to have shown, there is a good deal of flexibility in the hunting behavior of the large predators in general, such that neither the elephant-hunting lions of Chobe nor the buffalo-hunting lions of Lake Manyara offer a typical picture. If the Friesenhahn Cave mammoths were a prey specialty of *Homotherium*, then we may be seeing no more than a relatively localized phenomenon. Any depictions show simply part of a range of likely hunting tactics (plates 12 and 13).

TREATMENT OF THE KILL

Once a kill has been made, the priority among all predators is to consume the carcass without interference from others interested in a free meal, and to prevent the body from being stolen. For cats like the lion and the tiger, sheer size alone often acts as a sufficient deterrent, and a pride of lions does not really need to give way to any other predator. But just as lions are often able to steal carcasses from hyenas, so they may also lose their own kills to packs of hyenas and wild dogs, especially in cases where only one or two lions are involved.

One solution, adopted by many predators, is simply to eat as much food as possible as quickly as possible. Large predators are quite astonishingly speedy eaters with enormous capacities. While captive lions in a zoo may survive on perhaps 5 kg of meat per day for an adult male and 4 kg for a typical female, in the wild they take food when and where they find it and then eat as much as they can. Once a meal is commenced, a lion will usually eat until it is full; a male has been observed to eat 33 kg in one night, while four females have been seen to eat much of a male zebra in five hours. By that time the carcass is almost sure to have been discovered by other hungry animals, particularly hyenas, and a longer stay may be difficult. Here one can see the clear advantage of group hunting activity, which brings down large animals. It may be necessary to share, but such sharing can mean virtually complete and rapid consumption without the longer-term need to guard the kill in order to ensure future meals and the ability to drive off most scavengers in the meantime. Tigers, faced with much less competition for a carcass, tend to eat just as much in a given time, but they may also stay with the same body and eat from it for several days.

Cheetahs, defenseless in the face of all but the mildest of threats to steal from them, are also swift eaters, although they may eat only about 5–6 kg of meat per day. They usually take only one meal from a carcass, and they stop frequently to look around while doing so. But they frequently lose their prey, and at times to perhaps the least likely of competitors. Large male baboons are themselves equipped with fearsome canines, and the combined aggression of two or more of them at a kill site can be more than enough to drive off a lone cheetah or even a family group of mother and subadults. Such animals will readily scavenge from cheetah kills, and may serve as a ready indicator of how our earliest ancestors could have obtained meat on the African savannas two million or so years ago.

The most remarkable treatment of prey carcasses in an effort to prevent stealing is that undertaken by the leopard. For its size the animal is extremely strong, a character it demonstrates by its habit of caching the remains of its prey in trees where it is safe from prowling hyenas. It is quite possible for a leopard to carry a carcass in excess of its own body weight if necessary. In a tree it can eat in peace (figure 5.25), and it may consume up to 18 kg of meat in a twelve-hour period.

Determining whether other species in the past may have taken their prey up into the trees is of course not easy, although it would indeed have been a useful strategy (figure 5.26). But we can be reasonably sure, based on discoveries made by Bob Brain at the South African ape-man site of Swartkrans near Johannesburg, that leopards have done it for the past couple of million years. The bones found in the remains of the cave at the site, including those of the ape-men themselves, show a pattern of destruction very similar to that caused by leopard-sized cats. Leopards are the most

FIGURE 5.25 *A leopard carrying a deer carcass into a tree*
This illustration, set in the upper Pleistocene of Europe, shows a leopard taking the body of a roe deer, *Capreolus capreolus*, into a tree to protect it from the attention of a group of spotted hyenas. Once securely wedged in place, the carcass can be consumed at leisure.

FIGURE 5.26 Afrosmilus africanus *with a small suid carcass*

This picture illustrates many of the difficulties of reconstruction. We have depicted the small, early Miocene cat taking a pig carcass into a tree to escape from two large creodonts of the extinct *Megistotherium osteothlastes* at the Libyan locality of Gebl Zelten. Clearly, we have no direct evidence of such behavior. Moreover, the taxonomic position of the cat is uncertain: some authorities consider it a member of the genus *Metailurus*, while others place it in *Pseudaelurus*.

FIGURE 5.27 *Sketch of a leopard with a hominid carcass in a tree*

Caching prey in trees growing in the mouth of a sinkhole probably enabled leopards to escape the harassment of hyenas. As the carcass disarticulated, the bones would then fall into the opening, to be preserved in the sediments of the developing cave.

common larger predator in the deposit, and are therefore known to have been active in the area. In the region today, trees grow preferentially in the small pockets of soil trapped in the mouths of such caves, and the bones of any prey carried into such a tree would naturally drop to the ground and fall into the cave (figure 5.27).

For the saber-toothed cats, the large canines must have posed some constraints, since they would have affected the manner in which a carcass could be tackled and would certainly have restricted the ability to transport heavy items, as we have already mentioned in the case of the mammoths associated with *Homotherium* at Friesenhahn Cave in Texas. In the living large cats, the incisors, as we have seen, are relatively small, and the lower ones are set in a straight line between the canines. The upper incisors, however, are often set somewhat in front of the upper canines, so that they are used together with the canines for pulling at flesh and intestines. The carnassials can then be employed in a sideways bite to remove items prior to swallowing, and film of tigers eating shows them dealing with the meat on a carcass in precisely this way. In most of the saber-toothed species the larger canines would have somewhat restricted the ease with which the carnassials could be used in that manner, and would have been of little use themselves in detaching meat. But the fact that the canines fell some way behind the arc of the very much enlarged incisors would have meant that the incisors alone could be used to get at the edible portions of the carcass, and could probably have been used to tear items off and make them available to the carnassials in much the same way as in modern cats (figure 5.28). This feature of the anterior dentition is also seen in the genus *Dinofelis*, and most especially in *D. piveteaui*, the most dentally derived of that lineage.

FIGURE 5.28 Smilodon *biting with carnassials*
In order to tear mouthfuls of flesh from the body of a prey animal, living large cats often apply a lateral bite with the carnassials. In this kind of bite the gape of the jaws is not great, so that there is little or no clearance between the tips of even the relatively short upper and lower canines of modern cats. This suggests that the larger canines of saber-tooths would not have been a major hindrance to such activity. In living cats the walls of the mouth normally cover the carnassials in the resting position, but a slight contraction of the zygomaticus and/or depressor anguli oris muscles is enough to bare them, thus allowing the animal direct access to the meat with its slicing teeth. A similar condition probably existed in *Smilodon*, and the proportions of its skull suggest that it could comfortably apply the carnassial bite to carcasses as illustrated.

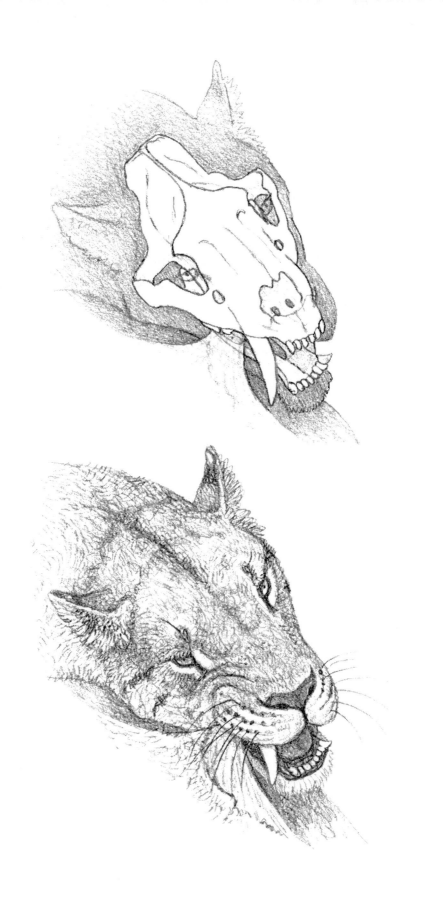

CHAPTER 6
The Changing Fauna

W E POINTED OUT EARLIER THAT MOST FORMS OF LIFE THAT have ever lived are now extinct. This extraordinary fact may take some time to grasp, but it remains true that the animals and many of the plants we see around us are as much newcomers on the face of the earth as we ourselves. Even over the relatively short span of time that humans or near-humans have been separated from our other primate relatives, some 5.0–7.0 Ma, great changes have occurred. If we multiply that figure by five to give us the life span of what we would recognize as the cat family, the changes have been enormous. Five million years ago, none of the living larger cat species was in existence. This is true wherever we look in the fossil record. In Africa at that time, saber-toothed cats and the so-called false saber-tooths of the genus *Dinofelis* were the only large felids. In Europe, Asia, and North America the same pattern is seen; South America and Australia, as has already been pointed out, had their own very particular faunas evolving in their own way.

These changes in the Felidae should be seen in their true perspective. In the first place, they are simply one feature of the massive change in the mammalian fauna that has taken place in the past several million years. If you looked at the larger mammalian fauna as a whole in those same regions at the Miocene-Pliocene boundary five million years ago, not a single species from the modern fauna would be found. In the second place, the evolution of the mammalian fauna in relatively recent times is itself not a unique feature of the fossil record. The dinosaurs were on the earth for many millions of years, but the various species also came and went over that time—so that a more detailed study of their fossil history would show the same kind of turnover.

Why have these enormous changes come about? Undoubtedly, accidents of history have played their part, and relatively local events such as volcanic eruptions, earthquakes, and other such natural occurrences will have determined some of the patterns that we see in the fossil record. Localized extinctions may have been driven by such events, and on islands it is evident that biotic factors such as the sudden incursion of invading species may have a detrimental effect on a native fauna or flora. But the most likely impetus to larger-scale evolutionary phenomena over long periods of time—that is, both extinctions *and* speciations, as well as changes in patterns of distribution—has surely been environmental change precipitated by climatic shifts, as argued persuasively by the South African paleontologist Elisabeth Vrba.

The climatic theory should be easy to substantiate because such an evolutionary mechanism would leave a clear signature in the fossil record: the major evolutionary events would be expected to clump together and correlate with significant changes in the physical environment. Moreover, environmental generalists would be expected to show a looser response to such perturbations in their environment. We therefore begin our discussion of faunal patterns by focusing on some of the major determinants of climate over the past few tens of millions of years.

CLIMATIC CHANGE

We tend to think of the climate as something fairly constant. The weather may change, but we see this as a temporary, at most seasonal, phenomenon. For most day-to-day purposes this is a reasonable point of view, but it is grossly misleading when we begin to look over periods spanning several millions of years. The real climatic pattern is complex, with variation as a major feature over geological time.

This variation has two major but interrelated causes. The first is the fact that the geography of the earth itself is changing, as the continents move and mountain chains build and erode (figure 6.1). Prior to about 100 Ma ago the continents of South America, Africa, Antarctica, and the Indian subcontinent formed one supercontinent known as Gondwanaland, with Eurasia in a second supercontinent known as Laurasia. The sections of Gondwanaland began to split up and move to their present positions on the globe around 100 Ma ago under the influence of plate tectonics, and these movements had profound effects on the structure of the oceans and on the temperature gradients, winds, and currents that were induced. One of the most important effects of these changes was that Antarctica moved to its present position at the South Pole, providing a landmass on which ice could begin to accumulate.

The second factor is the changes in solar energy received by the earth as a result of cyclical alterations in its orbit. These cycles have three major components (figure 6.2). The first is the variation in the shape of the orbit, which goes from more circular to more elliptical and then back, over a 100,000-year period: the more elliptical the orbit, the greater the contrast

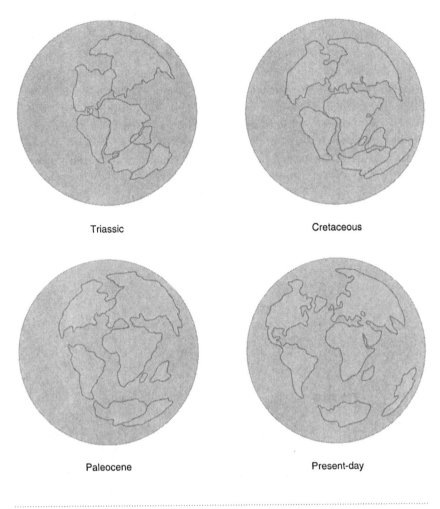

Triassic   Cretaceous

Paleocene   Present-day

FIGURE 6.1 *Movements of the continents*
The four drawings show the disposition of the continental landmasses in the
Triassic, Cretaceous, and Paleocene periods, as well as their present-day position.

in seasons in one hemisphere because the distance from the sun varies from
one season to the next. The second is the variation in the tilt of the rota-
tional axis of the earth, which fluctuates between 21.5 and 24.5 degrees in
a 41,000-year cycle: the greater the tilt, the greater the seasonal differences
in each hemisphere as that part of the planet leans either toward or away
from the sun. The third component is the precession, or wobble, of the
tilted axis of the earth's spin, which has a periodicity of 23,000 years. This
wobble determines at what point in the orbit, nearer to or farther from the
sun, the seasons fall. The simplest analogy is with a spinning top, one that
rotates not perfectly upright but with a slight and variable lean; when a top
starts to slow down, the axis of rotation may be seen to wobble in this way.

As the orbit changes, the energy received throughout the year rises and
falls. The climate will thus be affected in any circumstances, but geography
now begins to play an important part in the detailed pattern that emerges.
This is because with the continents placed as they are it becomes possible
for ice to build up at the southern polar region on Antarctica, on the land
within the Arctic Circle, and at high altitudes. This buildup occurs simply
because the summer temperatures are not high enough for a long enough
period each year to disperse the winter snowfall. Since the variations in
solar input are cyclical, the gross pattern of climatic change that they may

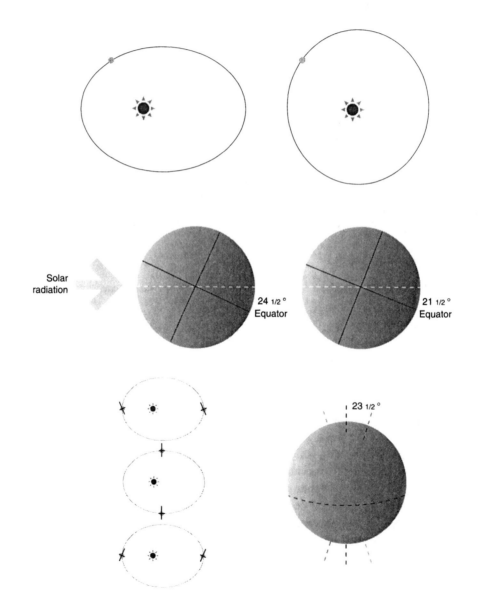

Solar
radiation

24 1/2 °
Equator

21 1/2 °
Equator

23 1/2 °

induce is itself cyclical. We have had numerous ice ages in the relatively recent past, approximately every 100,000 years over the past 800,000 years, during which time they have been most severe, and that pattern may be traced back further to perhaps 2.5 Ma ago. The interglacial periods (times when the ice has retreated) have been short—perhaps 10,000 years. We are currently in an interglacial, and it has lasted for around 10,000 years!

The point to reemphasize is that suitable combinations of continental disposition are the essential feature that tips the balance in favor of a glacial-interglacial cycle. The orbital variations are always there, but while they are essential they are not sufficient. Only with the continents in the right place can the land-based ice essential for the growth of significant ice sheets build up, and only then can the ocean currents play their part in the pattern of warming and cooling. If the continents had always been in their present positions we should have had ice ages every 100,000 years or so throughout the past—and we know that was not the case. There have been other periods of ice ages, but only when conditions were right.

We can see evidence of former ice sheets in the morphology of the landscape around us in many parts of the world, either in terms of the direct action of the ice itself as glaciers gouged their way across the terrain, or in terms of features left where the ground is frozen solid. But the evidence for the number of glacial cycles is naturally destroyed in land-based deposits, since later events obliterate earlier ones. The one place where this is not the case is in the ocean sediments, where the remains of marine organisms end up after death. The calcium carbonate of the shells of foraminifera (small sea creatures) is derived from sea water, and the composition of the shell reflects the temperature and salinity of the water at the time of its formation.

The precise trigger for each new glacial advance is not yet clear, but it may have something to do with the orbitally induced increase in seasonal

....

FIGURE 6.2 *Orbital changes*

The variations in the orbit of the earth interact to alter the amount of solar energy received over time.

Top: Eccentricity of the orbit. As the shape of the orbit varies, the seasonal contrast is enhanced or reduced.

Center: Tilt of the axis of rotation in relation to the plane of the orbit. This tilt produces the seasonal contrasts as each hemisphere points toward (summer) or away from (winter) the sun. As the tilt reduces, the polar regions receive less sunlight in summer.

Below right: Precession or wobble of the axis. This helps to produce the "precession of the equinoxes" (*below left*), the changes in the coincidence of summer and winter with the distance from the sun that results from the orbital shape. Thus if the Northern Hemisphere is tilted toward the sun (summer) at a time when the earth is farthest from the sun, the summers will be relatively colder.

contrasts and their effect upon the salinity and circulation of the ocean waters. The Northern Hemisphere is kept warm in winter by a massive northward current of very saline water in the Atlantic, the North Atlantic Drift, which loses its heat, sinks to the depths, and returns southward. Any disruption to that current would cause the Northern Hemisphere to cool significantly, reducing the summer temperatures and permitting ice to build up. Evidence from deep-sea cores suggests that just such a disruption to the current happened during glacial periods.

Just how much difference there has been between glacial and interglacial conditions at their peak may be shown by a few figures. Current estimates for the maximum ice advance of each glaciation suggest a lowering of the global sea level by as much as 130 meters, as the water is locked up in the land-based ice sheets, and a lowering of average July temperatures in Europe by around 10–15°C. The drop in average January temperature is of the order of 20°, which means that the permafrost and tundralike conditions that are now confined to higher latitudes then prevailed in much of northern Eurasia and North America. As the sea levels dropped, land became exposed around coastal margins, and land bridges existed between Britain and the continent of Europe across the channel, and between eastern Asia and western North America across the Bering Strait (figure 6.3).

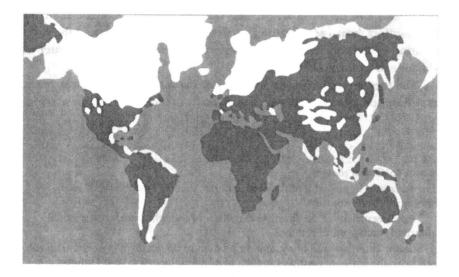

FIGURE 6.3 *The effects of ice sheets*
The figure illustrates typical ice sheet extensions during glacial maxima and the formation of land bridges, especially between Asia and North America: *white,* ice sheet; *black,* present-day land; *gray,* areas exposed by a fall in sea level during ice extension.

These are the extremes of the climatic cooling over the past million or so years. But over the past thirty million years during which the cats have evolved, a general pattern of global cooling may be seen, one that may be traced back to the time of the break-up of Gondwanaland and that resulted from the change in thermal gradients and the development of ocean currents.

THE PAST TEN MILLION YEARS

As we have seen, we can trace the history of the Felidae back into the Oligocene, around thirty million years ago. At that time the geography of the earth was still very different from today: South America, Antarctica, and Australia were completely isolated, and Africa had only tenuous connections with Europe.

During the upper Oligocene and early Miocene the dominant large carnivores of Eurasia were the amphicyonid bears (figure 6.4) and the hemicyonine protobears. The amphicyonids had a skeleton with a mixture of bearlike and catlike adaptations, while their dentition was clearly doglike. The solution to this anatomical puzzle may be that they were all-purpose carnivores, able to hunt actively, to scavenge, and to eat vegetable matter. They became extinct during the later Miocene, and their appearance at the Spanish site of Cerro Batallones is perhaps the latest. These animals were joined in the predator guilds of Eurasia and Africa by the giant hyenodont creodonts, enormous cursorial hunters and scavengers with oversized skulls of up to 60 cm in length that were the largest of any terrestrial meat-eating mammal.

Ecological associations of plants and animals began to change in a complex manner during the Oligocene, and those changes picked up in speed during the Miocene as grasslands opened up and grazing adaptations increased among the ungulates, leading among other things to an explosion in the diversity of ruminant families able to process vegetation more fully. Hypsodont (high-crowned) dentitions evolved, while skeletons adapted to running in more open terrain became more common. It is only in the earliest Miocene that members of the families Bovidae, Giraffidae, Cervidae, and Antilocapridae are known, for example. In contrast, the Persissodactyla (odd-toed ungulates today consisting of horses, rhinos, and tapirs) were at their most diverse during the Eocene and then declined in numbers, but they included small Miocene horses and rhinos that were often extremely large alongside chalicotheres. The latter were large, claw-hoofed animals of bizarre appearance by modern standards (plate 14), with limb proportions often said to resemble a gorilla and a probable life style that included pulling down vegetation while standing upright on the shortened hind legs. The increased emphasis on cursoriality among the horses and rhinos is matched by the extinction of the chalicotheres.

FIGURE 6.4 *A reconstruction of the amphicyonid bear dog,* Amphicyon major
This reconstruction is based on material from the lower Miocene site of Sansan in France. The animal was about the size of a modern brown bear and had powerful forelimbs somewhat resembling those of living bears, but its hind limbs and back were longer and somewhat catlike. The overall impression is of an animal more capable of leaping than any bear. The long, muscular tail could have aided in balancing during any such bounding, and it is easy to imagine that *Amphicyon major* was a rather active predator. The teeth resemble those of a wolf, and imply that its diet may also have been rather similar. These bear dogs became extinct during the upper Miocene.

The prey available to the earliest cats of the genera *Proailurus* therefore differed markedly from the modern fauna, allied to the fact that *Proailurus* itself was small and probably largely arboreal. It is only with *Pseudaelurus* that size increases led to the appearance of a range of animals up to the size of modern pumas, and that an increased emphasis on terrestrial locomotion can be seen in the skeleton.

But the vast majority of the felid species that we know are recorded from the last ten or so million years, and it is on this period that we shall concentrate. Within this period, we can identify perhaps three important, globally significant climatic events that stand out from the overall trend toward cooling. The first occurred between 6.5 and 5.0 Ma ago, at the end of the Miocene, when a major development in the Antarctic ice sheet and a lowering of global sea levels, in conjunction with some local uplift in the region of the Gibraltar Straits, led to the isolation of the Mediterranean. This isolation was followed by evaporation of the trapped waters, which led to a massive concentration of brine and to the deposition of enormous quantities of salt. The desiccated Mediterranean produced a dry-land connection between Europe and Africa, which lasted until around 5.0 Ma ago, and which would have enabled considerable movement between southwestern Europe and northern Africa.

The second event took place between 3.2 Ma and 2.5 Ma ago, and is first seen as the onset of glaciation in Iceland. Cycles of cooling have been identified in the pollen record of northern Europe at this time, and the typical summer drought of the Mediterranean region has been recorded. It was also at around this time that North and South America became permanently joined; although a geographic rather than a climatic event, this development had its own effect on the faunas of the two continents because major interchanges then became possible. By around 2.5 Ma ago we have evidence of the first major development of ice in the Northern Hemisphere. It is more difficult to produce a major northern ice sheet because there is no land at the North Pole itself, unlike the situation in the Antarctic, and it seems likely that the onset of glaciation in the Northern Hemisphere was linked to events at the South Pole via ocean currents. It is at this time that we see evidence of the development of the classic glacial-interglacial cycles. The developments at 2.5 Ma ago therefore seem to be a culmination of changes set in motion at 3.2 Ma ago. The third event, at around 900,000 years ago, was a downturn in the temperatures and the onset of the full glacial-interglacial pattern of climatic oscillation with its maximum swings.

All these events were marked to a greater or lesser extent by changes in the composition and distribution of the terrestrial mammal fauna—at times regionally, at other times globally. Of course the faunal changes that occurred, while driven ultimately by climate, depended for their precise timing and sequence of events upon the details of the faunal composition in each area. Land bridges came and went as the sea level fell and then rose again with changes in the ice sheets. The Beringian connection between Asia and North America is a particular case in point, and a host of North American species appear to have had their origins in the Old World. Gross environmental change, in other words, may be seen as a necessary condition for faunal change, but it is not always a sufficient one. The circumstances in which different populations of individual species found themselves will have varied depending on geography—particularly so in the case of the larger predators, with their tolerance for a variety of conditions provided that sufficient food be available. Thus while the saber-toothed cats became extinct in Africa at about 1.5 Ma ago, in Europe and Asia *Homotherium* lingered on for perhaps another million years, and in North America it only became extinct during the last glaciation (plate 15). In both North and South America, the more common *Smilodon* also became extinct only within the last 10,000–20,000 years.

Let us now look at some patterns of change in the mammalian faunas of various parts of the world against a background of the environmental changes of each of these three events.

The End-Miocene Event in Eurasia and North America

As we have said, the geography of Europe and of Asia changed considerably during the Miocene as a result of the major continental movements that produced the mountain chains of the Alps and the Himalayas, raised the Tibetan Plateau, and gradually closed up the seaways between the African, Eurasian, and Indian plates. The end result was to leave the Mediterranean as a landlocked sea, separated from the Indian Ocean, one that eventually dried up when crustal movements closed the portal at Gibraltar. Throughout the period, cooler winters, together with a decrease in summer rainfall, led to the gradual development of more open woodlands, changes no doubt correlated with the alterations in physical geography.

The final part of the Miocene, the period between 10.0 Ma and 5.0 Ma ago, is marked in Europe by a particular mammalian fauna usually referred to as the Turolian, named after the Teruel district of Spain. The Turolian may be seen as the final stage in the development of a more open-country fauna, but it should also be seen as simply one facet of a regionally diverse later Miocene mammalian assemblage. The mammalian fauna of the Greek island of Samos, for example, has been shown in an elegant study by Nikos Solounias and Beth Dawson-Saunders to include forest or woodland ruminants. Taken overall, the Eurasian fauna of the late Miocene was considerably more diverse than any modern savanna-dwelling community, especially in terms of the larger taxa such as proboscideans and giraffes, a point emphasized by Donald Savage and David Russell in their compilation of mammalian paleofaunas. Some typical elements of that fauna are depicted in figures 6.5 and 6.6. However, the beginnings of the Pliocene period after 5.0 Ma ago saw something of a swing back toward more forested conditions and a parallel shift in the composition of the mammalian fauna, generally accorded the name Ruscinian. This shift was of major importance: some 178 genera of land mammals are known from the Turolian, but 122 of these, 68%, do not appear in the earliest Pliocene. Indeed, some 13% of European land mammal families became extinct at that point.

Not unexpectedly, the Carnivora were also affected. In Europe, twelve hyena species and six species of cats were lost. Some of the cats were smaller animals, such as members of the genera *Paramachairodus* and *Metailurus*, but three were larger species of the genus *Machairodus*. Among the potential prey of these cats, and also suffering from the extinctions, were the family Bovidae (the antelopes) of the Turolian, reduced in numbers from around 60 species to perhaps 8 in the earliest Pliocene. We should see the changes in the composition of the antelopes, the cats, and the hyenas as very much interlinked effects of environmental change, not as a series of seemingly random and isolated events in the history of life.

For Asia, it is difficult to interpret the extent of true change between latest Miocene and earliest Pliocene because the number and distribution of fossil localities, to say nothing of the reliability of dates and identifications, are so variable. There would seem, on the face of it, to have been considerable change, perhaps of similar extent to that seen in Europe, with

FIGURE **6.5** *A sample of Turolian-age carnivores from the Mediterranean region*
From left to right: The large bear *Indarctos;* the small, doglike hyena *Hyaenotherium;*
the large, bone-cracking hyena *Adcrocuta;* the large felid *Machairodus;* and the
medium-sized felid *Paramachairodus.* Each square measures 50 cm on a side.

a marked impoverishment of the diversity in the Bovidae and other ungulate families.

In Africa at this time, while there is no equivalent of the European revolution in the fauna, there are indications of vegetational changes and a
marked wave of evolutionary activity in the Bovidae, with immigrations,
extinctions, and speciations. However, faunal change on a scale similar to
that in Europe was seen at the end of the Miocene in North America, where
74% of the genera and 18% of the families went extinct. Typical elements
of the Hemphilian fauna are shown in figures 6.7 and 6.8. American
changes in the felids are marked by the extinction of *Machairodus coloradensis* and by the Pliocene appearances of *Homotherium* and *Megantereon*,
and are paralleled by a reduction in the number of canids and the eventual
first appearance of the only hyena ever recorded in North America, the so-
called hunting hyena, *Chasmaporthetes ossifragus.* Among the prey species,
some of the major changes at the end of the North American Miocene
involve reductions in the number of horse species and, in the earliest
Pliocene, the first representatives of the genus *Equus,* to which modern
horses belong. North America at the time lacked the antelopes, but their
place was taken by a family known as the Antilocapridae, and for the larger
predators they would have made an equally attractive source of food.

The end of the Miocene is thus one of the most important points in the
evolution of the terrestrial mammalian fauna in general, and of the Carnivora in particular. In Eurasia we see a reduction in doglike morphotypes
among the Hyaenidae—that is, of animals with a more generalized dentition—and the appearance of animals with large, bone-smashing teeth. This
trend within the hyenas is eventually matched by a rise in numbers and in

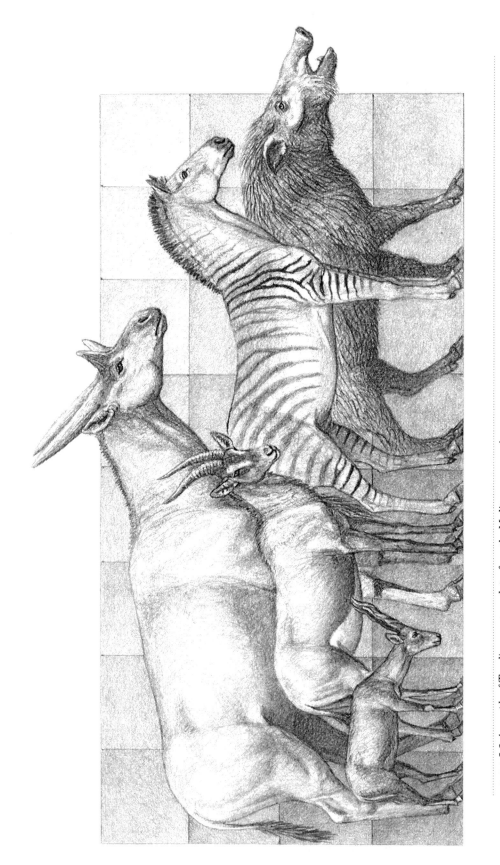

FIGURE 6.6 *A sample of Turolian-age ungulates from the Mediterranean region*

From left to right: The small antelope *Paleoreas*; the large giraffid *Birgerbohlinia*; the medium-sized antelope *Tragoportax*; the three-toed horse *Hipparion*; and the giant pig, *Microstonyx*. Each square measures 50 cm on a side.

FIGURE 6.7 *A sample of Hemphilian-age carnivores from North America*
From left to right: The bone-cracking borophagine canid *Osteoborus;* the ursid *Agriotherium;* the bone-cracking canid *Epicyon;* the nimravid *Barbourofelis;* and the large felid *Machairodus.* Each square measures 50 cm on a side.

FIGURE 6.8 *A sample of Hemphilian-age ungulates from North America*
From left to right: The pronghorn *Cosoryx;* the giraffelike camel *Aepycamelus;* the small three-toed horse *Nannipus;* the three-toed horse *Neohipparion;* and the peccary *Prosthennops.* Each square measures 50 cm on a side.

ecological importance of the Canidae themselves, a development that occurs somewhat later in Africa. It has been suggested by Larry Martin that social hunting by packs of animals such as the dogs may be an entirely Plio-Pleistocene development, and that the development of herding structure among herbivores in more open terrain is likely to have been one of the primary impetuses behind such a behavioral switch. Against this, however, we should point out that the more doglike hyenas of the Miocene may well have employed some form of group hunting, while the Miocene radiation of the ruminant artiodactyls—and especially the antelopes and deer, with their increasingly elaborate horns and antlers—may suggest herd structures of the kind likely to provoke cooperative behavior among carnivores.

What we therefore see in the earliest Pliocene is the emergence of a differently organized guild, although its roots can clearly be traced back into the Miocene. The large cats of the earliest Pliocene, all machairodonts, were largely specialized as flesh-eaters dependent on seizing prey and holding it still while killing it in order to avoid (or at least minimize) damage to their canines. The hyenas had started to show the specialized ability to demolish bone, and thus to obtain a consistent living from scavenging when necessary, although it is interesting to note that a second, major morphotype continued throughout the Pliocene in the form of the cursorial hunting hyenas of the genus *Chasmaporthetes*.

The Event Between 3.2 and 2.5 Ma Ago

In Europe, the beginning of this period of cooling is broadly correlated with the appearance of what is known as the Villafranchian fauna, which takes its name from the northwestern Italian locality of Villafranca d'Asti near Turin. The change from the preceding Ruscinian fauna is by no means absolute, and many of the forest species survive within what was evidently still a fairly wooded environment. True elephants are unknown at that time, although members of two related families of the order Proboscidea (long-nosed animals), Mammutidae and Gomphotheriidae, are found and exhibit the typical teeth of an animal that survives by browsing on leafy vegetation rather than by eating grasses. But the deer fauna of the Villafranchian is richer, with more advanced species of generally larger size, and new species of bovids and rhinos appear. The earlier and more primitive horses of the genus *Hipparion* are somewhat rare. Among the newcomers, however, are the European cheetah, the machairodonts *Homotherium latidens* and *Megantereon cultridens*, the hunting hyena, *Chasmaporthetes*, and perhaps the large *Pliocrocuta perrieri*, an animal first recorded in Asia at around 4.0 Ma ago. The latter was similar to a large brown hyena, and may have found a very good living by scavenging from the kills of the cats. *Chasmaporthetes lunensis*, in contrast, was a gracile animal rather like its close relative in North America, and it too may have provided an attractive source of carcasses for its scavenging companion.

Elements of the Villafranchian fauna of the Mediterranean are shown in figures 6.9 and 6.10.

The early part of this period in North America is not clearly represented in the fossil record, because many of the localities lack good dating evidence (although see below). But the land connection with South America that was established at about 3.2 Ma ago had important implications for the latter continent. In the period that followed a number of northern faunal elements began to appear there, including eventually the cats *Megantereon* and *Smilodon* and, later still, the jaguar, the puma, and even the lion. Movements northward are recorded for a variety of groups such as opossums, armadillos, and ground sloths, but not, it would seem, for any of the marsupial carnivores. On balance, the effect on the south was much greater than on the north, with more than 50% of the genera of modern-day South America descended from the invaders. By the late Pleistocene all the indigenous South American ungulates had become extinct, and it is difficult to avoid the conclusion that the incursion of placental carnivores played some part in that. To what extent the interchanges themselves reflect simple geographic opportunism as opposed to climatic impetus is currently unclear.

In Asia the problems of reliability continue to plague interpretation, but there is certainly a greater diversity of species known from the later part of the Pliocene, so much so that one suspects the earlier Pliocene impoverishment to be something of an artifact. However, a broadly Europe-like range of larger cats is known, with *Homotherium*, *Megantereon*, and *Acinonyx* clearly represented. There is also some evidence of interchange with North America, and it is perhaps at this time that *Megantereon* dispersed to the New World, possibly in conjunction with *Chasmaporthetes* and bears of the genus *Ursus*.

In Africa there is no particular faunal turnover at 3.2 Ma ago, and it may be that a sufficient mosaic of environments initially existed there to permit most species to survive any change in the environment. But from around at least 3.5 Ma ago a significant number of modern members of the guild are known in that continent. The lion, *Panthera leo*; the leopard, *Panthera pardus*; the cheetah, *Acinonyx jubatus*; and the spotted hyena, *Crocuta crocuta*, all appear in the fossil record of eastern, and later southern, Africa. They do not replace the older carnivores, however, because *Homotherium*, *Megantereon*, and *Dinofelis* all continue to be found in the same deposits as their more modern relatives.

Although the deep-sea core evidence for cooling shows that the change commencing at 3.2 Ma ago was an event of global significance, the effects of the climatic change are perhaps most apparent in the terrestrial mammalian fauna of Africa during the period between 3.0 and 2.0 Ma ago. This period saw the transformation of much of the closed forests and woodlands of eastern Africa into savannas, as well as the establishment of the Sahara

FIGURE 6.9 *A sample of Villafranchian-age carnivores from the Mediterranean region*

From left to right: The small, wolflike dog *Canis;* the scimitar-toothed cat *Homotherium;* the "hunting hyena," *Chasmaporthetes;* the giant cheetah, *Acinonyx;* the dirk-toothed cat *Megantereon;* and the hyena *Pliocrocuta.* Each square measures 50 cm on a side.

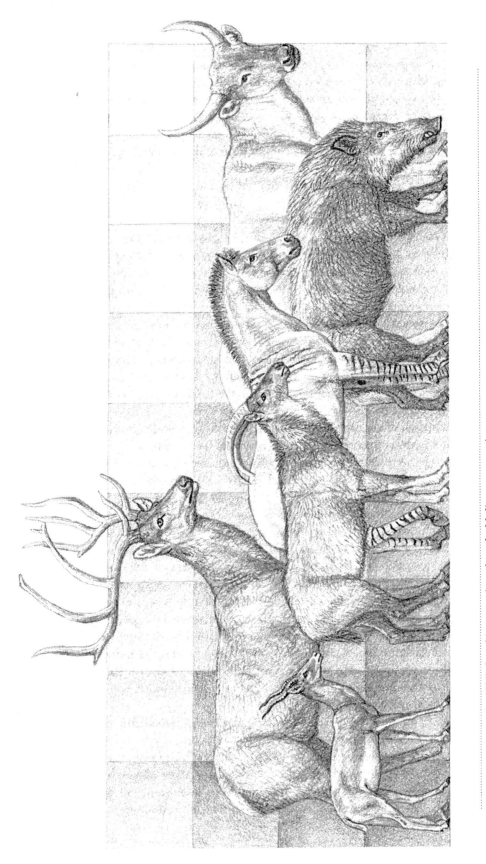

FIGURE 6.10 *A sample of Villafranchian ungulates from the Mediterranean region*

From left to right: The small antelope *Gazellospira*; the large deer *Eucladoceros*; the large rupicaprine *Gallogoral*; the small horse *Equus*; the wild boar *Sus*; and the primitive ox *Leptobos*. Each square measures 50 cm on a side.

desert as a permanent feature. Subsequent changes in the African carnivore guild appear to have been related to changes in the composition of the prey species population, in turn related to the major shift in conditions provoked by climatic shifts at 2.5 Ma ago when ice caps grew and the savannas spread. Between 2.5 and 2.0 Ma ago in particular, a whole range of changes may be seen both in the composition of the larger ungulate fauna and in the evolution of dentitions and other bodily features better suited to coping with tougher vegetation in more open and arid conditions.

The elephants are a case in point. Throughout the Pliocene of eastern Africa they show a trend toward extremely high-crowned and folded teeth with extensive enamel cutting ridges, but the rate of this change increases markedly in the later Pliocene, especially after around 2.3 Ma ago (figure 6.11). At that time the lineage represented by the now extinct *Elephas recki* became the dominant species, while the forerunners of the living African elephant genus, *Loxodonta*, declined. This pattern of massive dental development is seen in pigs as well, where several species underwent increases in the size of their teeth and in the height of the crown during the Pliocene and also showed an increase in the rate of change at this time. Horses of the living genus *Equus* first appear in Africa around 2.3 Ma ago, following a dispersion across Eurasia from the Americas via the Beringian land bridge, and the high-crowned teeth of these animals were well equipped for the consumption of abrasive fodder. The indigenous equids of the now extinct genus *Hipparion* coexisted with the newcomers for some time, and showed their own interesting pattern of dental development matching that of the pigs and elephants. At the same time, the white rhinoceros, *Ceratotherium simum*, exhibited changes in the structure of the skull, which became longer and enabled the species to graze on shorter grasses more easily.

Among the antelopes, the most taxonomically diverse of the African Plio-Pleistocene larger mammals, the dominance of more modern species in the savanna conditions points to the appearance of animals better adapted to running in open country. These changes in the structure of the antelope fauna are matched by changes in the dentition in later species of some lineages, such as a shift in emphasis to molars at the expense of premolars in alcelaphines like *Connochaetes*, the wildebeest. Such changes often involve the loss of the lower second premolar and the reduction of the third premolar to a peglike structure, an emphasis on the molar toothrow that seems to indicate increasing adaptation toward coping with more open grasslands.

The opening of the African vegetation and the increased dominance of antelopes capable of running faster suggest a major change of conditions for the larger carnivore guild, changes to which the saber-toothed cats may not have been quite so well adapted. Of course, as we have already seen, the living leopard, lion, and cheetah had long been in existence in Africa by that time, so that the subsequent changes in the composition of the carni-

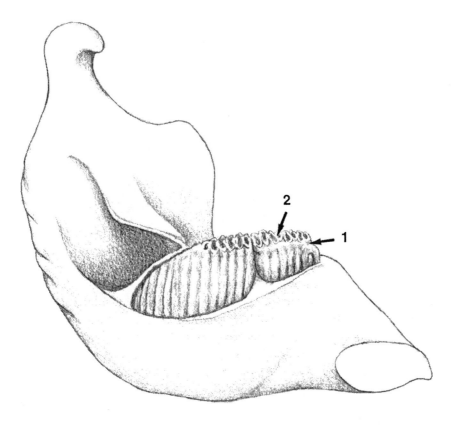

FIGURE 6.11 *Changing patterns of hypsodonty in* Elephas recki

The diagram shows a cutaway section of a typical left mandible of the eastern African Pliocene elephant *Elephas recki* from inside. Note the height of the crown of the large molar teeth (*1*), and the numerous, closely set loops of hard and very folded enamel set in cement (*2*). These loops wear more slowly than the dentine inside them, thus providing a constantly renewed cutting surface.

Over time, the height of the teeth, the number of enamel loops, and the extent of folding in the enamel increases, while the enamel itself becomes thinner. The effect is an increase in chewing capacity over the life span of the animal, and is most logically related to an increase in aridity and consequent changes in vegetation.

vore guild appear to stem from the extinction of the archaic saber-toothed and false saber-toothed species rather than from the origin of the modern cats. The two groups had coexisted in Africa for some 2.0 Ma before the archaic species went extinct, implying that simplistic ideas about competition as the mechanism for extinction require more thought.

In Europe at around 2.5 Ma ago, a change from a somewhat warmer climate with forests is also marked by the appearance of a fauna more characteristic of open landscapes. Chief among the members of this fauna are the first European elephants, with the appearance of the genus *Mammu-*

thus, the mammoth lineage, and the first European representatives of the advanced horse genus *Equus*, the latter as part of the dispersion that eventually took them to Africa. Mammoths, like horses, had high-crowned teeth, well equipped for dealing with the abrasive grasses that occur in more open environments, and in many respects the evolution of the mammoth lineage parallels that of the African *Elephas recki*. At the same time the archaic proboscideans become locally extinct and the more primitive horses of the genus *Hipparion* disappeared. The larger cats are represented in Europe at this time only by the cheetah and by the archaic, machairodont species *Homotherium latidens* and *Megantereon cultridens*, in a large carnivore guild virtually without dogs but still with the two large hyenas *Pliocrocuta perrieri* and *Chasmaporthetes lunensis*.

In North America there are indications of regional diversity in the vegetation at this period, with savanna conditions in the southwestern area and steppe vegetation in the west, although some interpretations suggest that the prairies of the central part of the continent were a later development. The larger cats consisted of four genera in a guild broadly similar to that of Europe. The American cheetah-like *Miracinonyx inexpectatus* was accompanied by at least one species of *Homotherium*, by *Megantereon*, and by a poorly known dinofelid species, *Dinofelis paleoonca*. The hyena *Chasmaporthetes* was also still present, but the only large bone-destroying scavenger was the so-called bone-eating dog, *Borophagus diversidens*, a member of the North American canid subfamily Borophaginae well known for its massive and partly hyena-like dentition.

The Event at 0.9 Ma Ago

This is a complicated period in the history of mammalian faunas, a time when changes set in train by earlier events receive something of a further jolt by a major shift in the frequency and intensity of climatic oscillations. The changes in the European fauna are heightened by the gradual rise to importance of the cold-tolerant species—animals such as the reindeer and bison, and eventually the woolly rhinoceros and the most extremely adapted of the mammoths. This period in Europe saw the disappearance of the dirk-toothed *Megantereon*, but the continued existence of *Homotherium* together with the cheetah, the European jaguar, and the first records in that continent of the lion, the leopard, and, perhaps a little later, the spotted hyena. The effect of the disappearance of *Megantereon* on the structure of the larger carnivore guild is likely to have added to the changes induced by the climate and by the appearance of more modern species. At around the same time, the ungulate fauna begins to record the appearance of larger and heavier-bodied species of deer and especially of bovids, the family to which cattle, bison, and sheep belong. Such heavier-bodied animals, even if no faster than the species that they replaced, would present a predator with new problems since success in predation is so clearly tied to prey size. It is perhaps significant that the smaller *Megantereon* became extinct at that point, possibly unable, despite its undoubted strength, to make a successful living in such circumstances. The larger, lion-sized *Homotherium* seems to have had no such immediate difficulties, and is in many respects one of the most successful of all the cats when one considers its longevity and dispersion.

Of course just as the environmental changes affected the vegetation and the animals that ate it, thus affecting the cats that preyed upon those animals, so the changes in the Felidae in turn had undoubted effects upon other species. The hyenas, once so varied, became reduced to just three major species during the relatively recent past: the spotted hyena, the brown hyena, and the striped hyena. In both Africa and Eurasia, the extinction of the saber-toothed cats is perhaps most notably accompanied by the extinction of one very large hyena, the so-called short-faced hyena, *Pachycrocuta brevirostris* (plate 16). This extraordinary beast was the size of a lion, with huge, strongly built jaws and massive teeth. Whether it hunted singly, or in packs like the spotted hyena of today, we simply do not know, although its bodily proportions suggest that it was not particularly fast moving. On the other hand, it was clearly equipped for scavenging, like all the larger hyenas, and was no doubt every bit as much an opportunist in the business of stealing carcasses as its modern relatives. The extinction of the saber-tooths would have affected the supply of carcasses, thus reducing the availability of food for scavengers.

The interactions between environment and fauna took different forms in different places and produced a variety of outcomes. In Europe and much of Asia the events of the previous several million years culminated for the purpose of our discussion here at around 0.5 Ma ago, when the last of the older species of larger carnivore became extinct. This brought to an end a period of overlap for the cats, during which living species such as the lion and leopard coexisted with the saber-toothed *Homotherium*. That modern guild structure, in turn, changed as we approach the present, doubtlessly aided by human interference. In America, a similar transition to the modern fauna took place, but at a much later date, with machairodont cats very much in evidence until some ten to twenty thousand years ago. Typical members of the American Rancho La Brean fauna are shown in figures 6.12 and 6.13. In Africa, by contrast, the transition occurred around 1.5 Ma ago, so that the African large predator guild has had a modern stamp for longer than anywhere else.

EXTINCTIONS

We should be clear about extinction. It has been an extremely common phenomenon in the development of life. That raises inevitable questions of why: Why did the machairodont cats go extinct? Why did *Homotherium* and *Megantereon*, two widely dispersed and therefore, by implication, successful

FIGURE 6.12 *Rancho La Brean carnivore guild*

From left to right: The dire wolf, *Canis dirus*; the saber-toothed cat *Smilodon*; the short-faced bear *Arctodus*; the cheetah-like cat *Miracinonyx*; and the lion, *Panthera leo* (*P. atrox* of some authors).

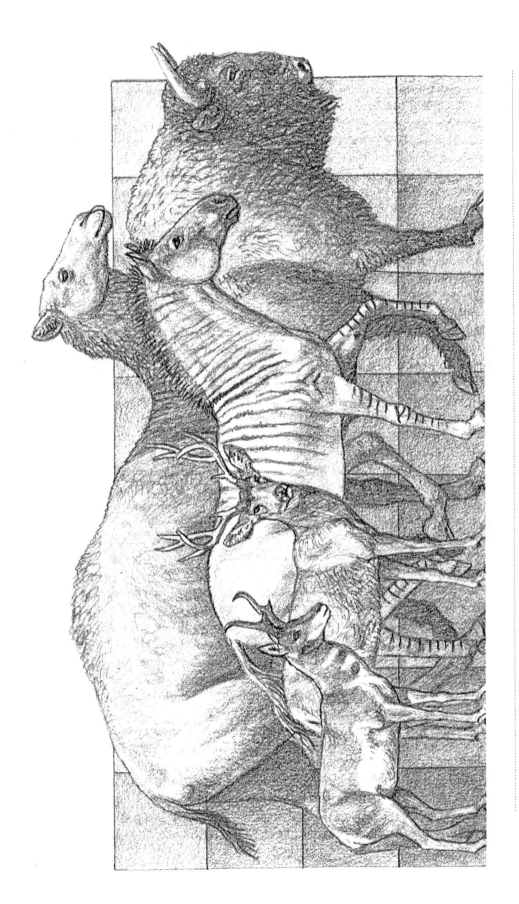

FIGURE 6.13 *Rancho La Brean ungulate guild*
From left to right. The pronghorn, *Antilocapra;* the deer *Odocoileus;* the giant camel, *Camelops;* the horse *Equus;* and the bison, *Bison.*

genera, disappear, and why did they do so first in Africa? And why did the machairodonts continue for so much longer in America, almost surviving into more recent, postglacial times? If the peculiar limb proportions of *Homotherium* appear to suggest some kind of adaptation to running in open terrain, then why did it become extinct in Africa at a time when the vegetation may have become more open, while surviving in North America and eventually being joined there by *Smilodon*, a probable descendant of *Megantereon*, which developed similarly unusual limb proportions?

Frankly, it is always difficult to answer such questions fully, but it is particularly hard to do so if we confine our attention to any one particular species or event. The fossil record is most useful for answering questions about general patterns and for looking at larger questions than simply the extinction or the origin of a single species. Thus, if we consider the time difference between the American and African extinction of the machairodonts we should also bear in mind the fact that the wider circumstances were different in the two continents. Africa sits astride the equator and has been in more or less its present position since the Pliocene (although there has been some northward movement of about 200 km over the past 4.0 Ma), while North America sits outside the tropics. The initial early Pliocene patterns of climatic variation and vegetation in the two were therefore very different, and thus the pattern of response to a global climatic event also differed, as we have seen.

Beyond that, the ungulate faunas of the two continents, and hence the type of prey available to the larger predators, also differed. The Americas lacked the Bovidae, the family of the African antelopes, until well into the Pleistocene, and even then the fauna was nothing like as extensive as that of Africa. The New World also had virtually no cervids until around the same time, and the only broadly equivalent family, the Antilocapridae, was represented by relatively few species during the Pliocene. Now it is true that a large carnivore is little concerned with the niceties of taxonomy, and cares not at all whether its prey is an antelope, a cervid, or anything else as long as it can be caught and provides a meal. But the different families of ungulates have different patterns of variation in body size and proportions, and with such differences go differences in locomotion and predator-avoidance strategy. The evolutionary history of such families also differs, and the response to any given selective pressures will therefore also have varied. America without either antelopes or cervids had few species sufficiently similar to the African antelopes in terms of the prey presented to its large predators, or in terms of a fauna likely to exhibit a similar response to a given climatic change. The climatic changes after 3.2 Ma ago, which altered the structure of the African ungulate fauna (and of the antelopes in particular) and made cursorial species more common, could not have the same effect on an American fauna that lacked antelopes. Given such differences it is hardly surprising that the response of the predator guild in each continent should also have varied. Moreover, those guilds themselves, and hence the interactions within them, differed: America had only one hyena, the gracile, hunting species *Chasmaporthetes ossifragus*, which became extinct some time after 2.0 Ma ago, and only *Panthera onca* and *Miracinonyx* among the larger feline cats until the arrival of the lion.

At the end of the Pleistocene a wholesale change took place in many of the mammalian faunas of the world, leading to the extinction in America of the saber-tooths as well as the lion and *Miracinonyx*. In Europe and Asia the lion and leopard became extinct over much of their territory, as did the spotted hyena, but these changes also affected the larger prey species in both continents as well. In Australia, long separated from the rest of the world, a different but no less drastic extinction of many faunal elements can also be seen. Only Africa seems to have escaped with its fauna largely intact. These end-Pleistocene extinctions have been the subject of intensive and extensive discussion, with climate and human intervention as perhaps the most frequently invoked explanations, but the problem remains unresolved. The possible involvement of human activity is perhaps the main reason for this uncertainty, for it is difficult to remove this factor from any consideration of underlying mechanisms.

If extinction has been common in the past then it is perhaps only to be expected that it will continue at the present day. But we should be equally clear about the fact that extinctions are *caused*. They do not happen randomly, or because a species has somehow run out of evolutionary momentum. This idea has been commonly put forward in the literature, specialist as well as popular, as an explanation for the extinction of all manner of organisms. Favorite among these have been the saber-toothed species, animals seen as developing inexorably along a fixed pathway until they became quite unsuited to their circumstances. However, some extinctions may simply have occurred as a response to relatively short-term circumstances and may well have been readily reversible. The extinction of horses in America at the end of the Pleistocene is a case in point. Whatever the precise cause of that extinction, there was nothing operating to prevent the domestic horses that escaped following reintroduction by Europeans from thriving in the very regions where their ancestors once roamed. In contrast, the living cheetah, which appears so superbly adapted to its African environment, may very well have teetered on the edge of extinction only a few thousand years ago, as evidenced by the extreme genetic uniformity of all populations. If there was indeed a "genetic bottleneck" for the cheetah, then they are with us almost through pure chance, and the border between success or failure may be very narrow indeed. The implication is that some of the extinct saber-toothed species might survive perfectly well in a suitable modern setting.

The idea of extinction as an inevitable response to overspecialization is therefore in many ways an oversimplification. It is certainly true that if organisms are unable to cope with changed circumstances they are extremely likely to become extinct—but the point to emphasize is the change in the circumstances, not the inevitability of the extinction itself. Neither the saber-toothed cats nor any other species that we have dealt with here went extinct in a vacuum, disappeared in a unique event unrelated to other changes taking place about them. But that said, we humans are now often *the* most important factor in the physical and biotic environment of many organisms, able to affect survival by a chance or deliberate action. This is particularly the case for the large predators, animals whose lifestyle fits badly with the requirements of an ever-expanding and increasingly polluting human population. It is probably true to say that most large predators in the wild are under some threat of extinction, and this is especially true of the larger cats. Hunted because of their depredations on livestock, for their skins, and simply for the "sport" of killing them, tigers, cheetahs, leopards, jaguars, and perhaps above all snow leopards are especially at risk, and ultimately may be visible only in a cage (figure 6.14).

THE EVOLUTION OF THE LARGE CATS: AN OVERVIEW

Several overall trends can be seen in the evolution of the large cats. First, we see a considerable increase in size and diversity over the thirty or so million years of their development. The first members of the Felidae were quite small, forest-living animals, confined to Eurasia and very similar in many ways to the living genets. *Proailurus* had a greater number of teeth than living cats, with short limbs and a plantigrade stance, suggesting that it was quite at home in the trees. *Pseudaelurus,* the likely descendent genus, begins to show the pattern of diversity emerging, with lineages leading to the saber-toothed machairodonts and the conical-toothed cats of today and a range of sizes from domestic-catlike to pumalike. But it is interesting to note that the living cats appear to have remained relatively small for much of that subsequent history, and only during the past 5.0 Ma do we see the appearance of the very large pantherines such as the lion and the tiger. Some of the saber-toothed cats in the meantime achieved enormous size, with *Machairodus* rivaling the largest of the living species. The machairodonts also developed an intriguing range of body plans, with the robust and relatively short-limbed *Smilodon* and *Megantereon* contrasting oddly with the long-legged and somewhat gracile *Homotherium.* The long forelimbs of the latter give the animal a particularly unusual appearance, somewhat reminiscent of a hyena—a trend seen also in the eastern South American species *Smilodon populator.* However, all the machairodonts seem to have shared the need for considerable strength in their forequarters and relatively shortened backs to enable them to bring down and hold prey firmly

FIGURE 6.14 *Portrait of a black leopard behind bars*

Such a scene may be the only view that future generations will ever have of a large cat. Although great efforts are being made to breed endangered cat species in captivity, the costs and difficulties of reintroduction programs are enormous. If habitat destruction cannot be stopped it is unlikely that we shall be able to recover what is lost, and in any event local inhabitants tend to see reintroduced predators as an unwelcome danger to life and property. The protection and management of free-living large cats and their environment today offer the only realistic hope of having these animals in a natural setting in the future.

so that sufficient damage might be inflicted by the teeth without undue risk to the long and relatively fragile upper canines.

The appearance of saber-toothed dentitions among the Felidae offers an interesting parallel with similar evolutionary developments in other groups. The most notable of these other groups is the catlike family Nimravidae, in which enormous upper canines evolved in a range of powerfully built animals with similarly strong forelimbs and large, retractable claws. Such close parallels enable us to see the saber-toothed cats in a comparative light, and amply demonstrate that theirs was by no means an unusual and pointless evolutionary track. They may ultimately have gone extinct, but they are hardly unusual in suffering that fate, and the development of long upper canines was quite clearly a successful strategy across a wide range of species.

So far as the behavior of fossil cats is concerned, the general features of lifestyle, social activity, and prey hunting and dispatch that we may observe in living animals allow us to make some overall assessments based on skeletal evidence. The number of skeletons of *Smilodon* found in the Rancho La Brea tar deposits, together with the evidence of healed injuries, imply a social structure that would enable sick or injured animals to survive hard times, while the footprints of the ancestral *Pseudaelurus* point to family group structures among even the earliest cats.

The skeletons and dentitions of the conical-toothed cats, living and fossil, betray a range of lifestyles and hunting and killing strategies distinct from those of the machairodonts. The most extreme of these are the highly cursorial adaptations of the cheetahs and of the American genus *Miracinonyx*. If the latter are not particularly closely related to the Old World, then the degree of parallel evolution is striking. The giant cheetah of the European Pleistocene, *Acinonyx pardinensis*, must have extended the range of prey sizes accessible to such a fast-running predator.

WHERE TO SEE FOSSIL CATS

Large museums, especially in countries with either a colonial history or their own native species, often have some material of living cats, sometimes mounted as complete skeletons for display. In contrast, original fossil specimens are obviously relatively rare, but many museums of natural history have casts of specimens from a variety of sources, and skulls of the American saber-tooth *Smilodon* are particularly popular. However, museum exhibitions tend to come and go, and it is no longer fashionable to have so many items on display as it once was. As a result, it is by no means easy to see fossil cats or nimravids everywhere in public galleries, or to guarantee that given specimens will be on show. It is therefore always worth asking at nearby museums, especially about material that may be kept in storage. The following short guide, based on institutions known to us, should be of some value in tracking down material.

- ARGENTINA: The Bernardino Ribadavia Museum of Buenos Aires has a complete skeleton of *Smilodon populator*.

- FINLAND: The Zoological Museum in Helsinki has a life-sized reconstruction of *Homotherium*.

- FRANCE: The Museum of Natural History in Paris has various items, including skeletons of European "cave lions," *Panthera spelaea*; skulls of *Homotherium* and *Megantereon*; and a skeleton of *Smilodon*.

In Lyon the Museum of Natural History has skulls of the European cheetah, *Acinonyx pardinensis*, while the Department of Earth Sciences at the Université Claude Bernard has the skeleton of *Homotherium* from Senèze.

• ITALY: The museum in the department of Earth Sciences of the University of Florence has a variety of Villafranchian fossil specimens, which include some rather fragmentary felid skulls and an interesting collection of leopard remains from last glaciation deposits in Equi Cave.

• SOUTH AFRICA: The Transvaal Museum in Pretoria has remains from several of the Transvaal hominid sites, including very fine skulls of *Dinofelis piveteaui* and *Dinofelis barlowi*.

The South African Museum in Cape Town has excellent specimens from the Mio-Pliocene locality of Langebaanweg, including an interesting series of *Dinofelis* mandibles.

• SPAIN: The Museum of Natural Sciences in Madrid has skeletons of *Machairodus* and *Paramachairodus*.

The Archaeological Museum in Banyoles has several very fine skulls of *Homotherium*.

• SWITZERLAND: The Natural History Museum in Basel has a mounted skeleton of *Megantereon cultridens*.

• UNITED KINGDOM: The Natural History Museum in London has extensive collections of specimens that include felid remains from many parts of the world.

• UNITED STATES OF AMERICA: The G. C. Page Museum of Los Angeles displays mounted skeletons of *Smilodon fatalis*.

The Los Angeles County Museum displays mounted skeletons of the Oligocene nimravids *Nimravus* and *Hoplophoneus*.

The Texas Memorial Museum has a mounted skeleton of *Homotherium serum* from Friesenhahn Cave.

The Florida State Museum has a mounted skeleton of the nimravid *Barbourofelis lovei*.

The National Museum of Natural History in Washington has mounted skeletons of *Hoplophoneus* and *Smilodon fatalis*.

The American Museum of Natural History in New York has skeletons of *Smilodon populator*, *Smilodon fatalis*, and *Hoplophoneus mentalis*.

EPILOGUE

THE EVOLUTION OF THE LARGE CATS HAS PRODUCED A SERIES OF fully functional forms, each with its own beauty of line and action. But the admiration that we may feel when confronted with even the best and most complete fossil is often tinged with frustration when we realize that so much about these animals will never really be known. No matter how intensely we study the bones of the saber-toothed cats, we will never see them actually moving and stalking their prey. Our guide to the range of their behavior will therefore have to be their living relatives.

However, although those who study living animals may seem to have all the information that they need, the difference between the paleontologist and the zoologist is really one of degree. Once the basic discovery of a species has been documented, science tends to ask complex questions that go beyond the study of the living animal itself. Such questions involve the place and role of individual species in the ecosystem, and when we ask them we begin to realize how far we are from knowing all the answers about even the most common living animals. For many it may of course already be too late to find out more. Entire species of large cats, such as the tiger or the snow leopard, may simply go extinct in the wild in our own lifetimes, and even surviving populations often live in such disturbed conditions that their real behavior is masked beyond recognition.

The imminent danger of humanly induced extinction for large cats would be not only a scientific loss. As George Schaller has said of the tiger, in the film *Tiger Crisis,* "future generations would feel truly sad that we in the twentieth century had such lack of foresight, such lack of compassion, so little sense of future, as to annihilate one of the most beautiful and powerful species that this planet has ever seen." All of us who enjoy the beauty of the large cats, and have any curiosity about their true nature and their evolution, owe a debt to those who are fighting to preserve what is left of wild populations and their habitats. For those involved in conservation efforts, human greed and ignorance often mean not only obstacles but also very real dangers. We should also remember that our own way of life is often the very force driving wild animals to extinction. It is our shared responsibility to ensure that extinction does not occur.

FURTHER READING

Anatomy, Locomotion, and Function

Akersten, W. A. 1985. Canine function in *Smilodon* (Mammalia, Felidae, Machairodontinae). *Los Angeles County Museum Contributions in Science* 356:1–22.

Emerson, S. B. and L. Radinsky. 1980. Functional analysis of saber-tooth cranial morphology. *Paleobiology* 6:295–312.

Gambaryan, P. P. 1974. *How Mammals Run: Anatomical Adaptations.* New York: Wiley.

Gonyea, W. J. 1976. Adaptive differences in the body proportions of large felids. *Acta Anatomica* 96:81–96.

——. 1976. Behavioral implications of saber-toothed felid morphology. *Paleobiology* 2:332–342.

Gonyea, W. J. and R. Ashworth. 1975. The form and function of retractile claws in the Felidae and other representative carnivorans. *Journal of Morphology* 145:229–238.

Hemmer, H. 1978. Socialization by intelligence: Social behavior in carnivores as a function of relative brain size and environment. *Carnivore* 1:102–105.

Hildebrand, M. 1959. Motions of the running cheetah and horse. *Journal of Mammalogy* 40:481–495.

——. 1961. Further studies on the locomotion of the cheetah. *Journal of Mammalogy* 42:84–91.

Hildebrand, M., D. M. Bramble, K. F. Liem, and D. B. Wake. 1985. *Functional Vertebrate Morphology.* Cambridge: Harvard University Press.

Miller, G. J. 1969. A new hypothesis to explain the method of food ingestion used by *Smilodon californicus* Bovard. *Tebiwa* 12:9–19.

——. 1980. Some new evidence in support of the stabbing hypothesis for *Smilodon californicus* Bovard. *Carnivore* 3:8–19.

——. 1984. On the jaw mechanism of *Smilodon californicus* Bovard and some other carnivores. *IVC Museum Society Occasional Paper* 7:1–107.

Peters, G. and M. H. Hast. 1994. Hyoid structure, laryngeal anatomy, and vocalization in felids (Mammalia: Carnivora: Felidae). *Zeitschrift für Säugetierkunde* 59:87–104.

Radinsky, L. 1975. Evolution of the felid brain. *Brain, Behavior and Evolution* 11:214–254.

Rawn-Schatzinger, V. 1983. Development and eruption sequence of deciduous and permanent teeth in the saber-tooth cat *Homotherium serum* Cope. *Journal of Vertebrate Paleontology* 3:49–57.

Robinson, R. 1978. Homologous coat color variation in *Felis*. *Carnivore* 1:68–71.

Taylor, M. E. 1989. Locomotor adaptation by carnivores. In J. L. Gittleman, ed., *Carnivore Behavior, Ecology, and Evolution*, pp. 382–409. New York: Comstock-Cornell.

Tejada-Flores, A. E. and C. A. Shaw. 1984. Tooth replacement and skull growth in *Smilodon* from Rancho La Brea. *Journal of Vertebrate Paleontology* 4:114–121.

Turnbull, W. D. 1978. Another look at dental specialization in the extinct sabre-toothed marsupial, *Thylacosmilus*, compared with its placental counterparts. In P. M. Butler and K. E. Joysey, eds., *Development, Function and Evolution of Teeth*, pp. 339–414. London: Academic Press.

EVOLUTION

Ridley, M. 1993. *Evolution*. Oxford: Blackwell Scientific Publications.

Turner, A. 1993. Species and speciation: Evolution and the fossil record. *Quaternary International* 19:5–8.

Turner, A. and H. E. H. Paterson. 1991. Species and speciation: Evolutionary tempo and mode in the fossil record reconsidered. *Geobios* 24:761–769.

Vrba, E. S. 1985. Environment and evolution: Alternative causes of the temporal distribution of evolutionary events. *South African Journal of Science* 81:229–236.

——. 1987. Ecology in relation to speciation rates: Some case histories of Miocene-Recent mammal clades. *Evolutionary Ecology* 1:283–300.

——. 1992. Mammals as a key to evolutionary theory. *Journal of Mammalogy* 73:1–28.

EXTINCTION

Martin, P. S. and R. G. Klein, eds. 1984. *Quaternary Extinctions: A Prehistoric Revolution*. Tucson: University of Arizona Press.

Martin, P. S. and H. E. Wright, eds. 1967. *Pleistocene Extinctions: The Search for a Cause*. New Haven: Yale University Press.

Owen-Smith, N. 1987. Pleistocene extinctions: The pivotal role of megaherbivores. *Paleobiology* 13:351–362.

Stuart, A. J. 1991. Mammalian extinctions in the late Pleistocene of northern Eurasia and North America. *Biological Review* 66:453–562.

FAUNAL EVOLUTION

Carroll, R. L. 1988. *Vertebrate Paleontology and Evolution*. New York: Freeman.

Flynn, L. J., R. H. Tedford, and X. Qiu. 1991. Enrichment and stability in the Pliocene mammalian fauna of North China. *Paleobiology* 17:246–265.

Kurtén, B. 1968. *Pleistocene Mammals of Europe.* London: Weidenfeld and Nicholson.

Kurtén, B. and E. Anderson. 1980. *Pleistocene Mammals of North America.* New York: Columbia University Press.

Lindsay, E. H., V. Fahlbusch, and P. Mein, eds. 1990. *European Neogene Mammal Chronology.* New York: Plenum Press.

Maglio, V. J. and H. B. S. Cooke. 1978. *Evolution of African Mammals.* Cambridge: Harvard University Press.

Savage, D. E. and D. E. Russell. 1983. *Mammalian Paleofaunas of the World.* London: Addison-Wesley.

Savage, R. J. G. and M. R. Long. 1986. *Mammal Evolution.* London: British Museum (Natural History).

Stuart, A. J. 1982. *Pleistocene Vertebrates in the British Isles.* London: Longman.

Turner, A. 1990. The evolution of the guild of larger terrestrial carnivores during the Plio-Pleistocene in Africa. *Geobios* 23:349–368.

——. 1992. Villafranchian-Galerian larger carnivores of Europe: Dispersions and extinctions. In W. von Koenigswald and L. Werdelin, eds., *Mammalian Migration and Dispersal Events in the European Quaternary*, pp. 153–160. *Courier Forschungsinstitut Senckenberg* 153.

Turner, A. and B. A. Wood. 1993. Taxonomic and geographic diversity in robust australopithecines and other African Plio-Pleistocene mammals. *Journal of Human Evolution* 24:147–168.

——. 1993. Comparative palaeontological context for the evolution of the early hominid masticatory system. *Journal of Human Evolution* 24:301–318.

Walter, G. H. and H. E. H. Paterson. 1994. The implications of palaeontological evidence for theories of ecological communities and species richness. *Australian Review of Ecology* 19:241–250.

FOSSIL FORMATION AND RECOVERY

Andrews, P. 1990. *Owls, Caves and Fossils.* London: British Museum (Natural History).

Behrensmeyer, A. K. and A. P. Hill, eds. 1980. *Fossils in the Making.* Chicago: University of Chicago Press.

Brain, C. K. 1981. *The Hunters or the Hunted?* Chicago: University of Chicago Press.

Wang, X. and L. D. Martin. 1993. Natural Trap Cave. *National Geographic Research and Exploration* 9:422–435.

FOSSIL SPECIES

Adams, D. B. 1979. The cheetah: Native American. *Science* 205:1155–1158.

Ballesio, R. 1963. Monographie d'un *Machairodus* du gisement de Senèze: *Homotherium crenatidens* Fabrini. *Traveaux de la Laboratoire de Géologie de l'Université de Lyon* 9:1–129.

Baskin, J. A. 1984. Carnivora from the late Clarendonian Love Bone Bed, Alachua County, Florida. Ph.D. diss., University of Florida.

Beaumont, G. de. 1975. Recherches sur les félidés (mammifères, carnivores) du Pliocène inférieur des sables à *Dinotherium* des environs d'Eppelsheim (Rheinhessen). *Archives des Sciences* 28:369–405.

——. 1978. Notes complémentaires sur quelques félidés (carnivores). *Archives des Sciences* 31:219–227.

Belinchón, M. and J. Morales. 1989. Los carnívoros del Mioceno inferior de Buñol (Valencia, España). *Revista Española de Paleontología* 4:3–8.

Berta, A. 1985. The status of *Smilodon* in North and South America. *Los Angeles County Museum Contributions in Science* 370:1–15.

——. 1987. The sabercat *Smilodon gracilis* from Florida and a discussion of its relationships (Mammalia, Felidae, Smilodontini). *Bulletin of the Florida State Museum of Biological Sciences* 31:1–63.

Boule, M. 1906. Les grands chats des cavernes. *Annales de Paléontologie* 1:69–95.

Bryant, H. N. 1988. Delayed eruption of the deciduous upper canine in the saber-toothed carnivore *Barbourofelis lovei* (Carnivora, Nimravidae). *Journal of Vertebrate Paleontology* 8:295–306.

Churcher, C. S. 1966. The affinities of *Dinobastis serus* Cope 1893. *Quaternaria* 8:263–275.

——. 1984. The status of *Smilodontopis* (Brown, 1908) and *Ischyrosmilus* (Merriam, 1918). *Royal Ontario Museum of Life Sciences Contribution* 140:1–59.

Cooke, H. B. S. 1991. *Dinofelis barlowi* (Mammalia, Carnivora, Felidae) cranial material from Bolt's Farm, collected by the University of California African Expedition. *Palaeontologia Africana* 28:9–21.

Croizet, J. B. and A. C. G. Jobert. 1828. *Recherches sur les Ossemens Fossiles du Département du Puy-de-Dôme*. Paris.

Crusafont Pairó, M. and E. Aguirre. 1972. *Stenailurus*, félidé nouveau, du Turolien d'Espagne. *Annales de Paléontologie, Vertébrés* 58:211–223.

Dawkins, W. B. and W. A. Sandford. 1866–1872. *A Monograph of the British Pleistocene Mammalia*. Vol. 1, *British Pleistocene Felidae*. London: Palaeontographical Society.

de Bonis, G. 1976. Un félidé à longues canines de la colline de Perrier (Puy-de-Dôme): Ses rapports avec les félinés machairodontes. *Annales de Paléontologie* 62:159–198.

Evans, G. L. 1961. The Friesenhahn Cave. *Texas Memorial Museum Bulletin* 2:3–22.

Ficcarelli, G. 1978. The Villafranchian machairodonts of Tuscany. *Palaeontographia Italica* 71:17–26.

——. 1984. The Villafranchian cheetahs from Tuscany and remarks on the dispersal and evolution of the genus *Acinonyx*. *Palaeontographia Italica* 73:94–103.

Harrison, J. A. 1983. Carnivora of the Edson local fauna (Late Hemphilian), Kansas. *Smithsonian Contributions to Paleobiology* 54.

Hemmer, H. 1978. Considerations on sociality in fossil carnivores. *Carnivore* 1:105–107.

Hendey, B. 1974. The late Cenozoic Carnivora of the south-western Cape Province. *Annals of the South African Museum* 63:1–369.

Hibbard, C. W. 1934. Two new genera of Felidae from the Middle Pliocene of Kansas. *Transactions of the Kansas Academy of Sciences* 37:239–255.

Hooijer, D. A. 1947. Pleistocene remains of *Panthera tigris* (Linnaeus) subspecies from Wanhsien, Szechwan, China, compared with fossil and recent tigers from other localities. *American Museum Novitates* 1346:1–17.

Hunt, R. 1987. Evolution of the Aeluroid Carnivora: Significance of auditory structure in the nimravid cat *Dinictis*. *American Museum Novitates* 2886:1–74.

Koufos, G. D. 1992. The Pleistocene carnivores of the Mygdonia Basin (Macedonia, Greece). *Annales de Paléontologie* 78:205–259.

Kurtén, B. 1965. The Pleistocene Felidae of Florida. *Bulletin of the Florida State Museum* 9:215–273.

——. 1973. The genus *Dinofelis* (Carnivora, Mammalia) in the Blancan of North America. *Pearce-Sellards Series* 19:1–7.

——. 1973. Pleistocene jaguars in North America. *Commentationes Biologicae* 62:1–23.

——. 1976. Fossil puma (Mammalia: Felidae) in North America. *Netherlands Journal of Zoology* 26:502–534.

——. 1985. The Pleistocene lion of Beringia. *Annales Zoologici Fennici* 22:117–121.

Kurtén, B. and L. Werdelin. 1990. Relationships between North and South American *Smilodon*. *Journal of Vertebrate Paleontology* 10 (2): 158–169.

Lund, P. W. 1950. *Memorias sóbre a Paleontología Brasileira Revistas e Comentadas por Carlos de Paula Couto*. Río de Janeiro: Ministerio de Educao e Saude Instituto Nacional do Livro.

Lydekker, R. B. A. 1884. Indian Tertiary and post-Tertiary Vertebrata. *Palaeontologia Indica* 2:1–363.

Marean, C. W. and C. L. Ehrhardt. 1995. Paleoanthropological and paleoecological implications of the taphonomy of a sabretooth's den. *Journal of Human Evolution* 29:515–547.

Merriam, J. C. and C. Stock. 1932. The Felidae of Rancho La Brea. *Carnegie Institution of Washington Publications* 442:1–231.

Miller, G. J. 1968. On the age distribution of *Smilodon californicus* Bovard from Rancho La Brea. *Los Angeles County Museum Contributions in Science* 131:1–17.

Petter, G. and F. C. Howell. 1987. *Machairodus africanus* Arambourg, 1970 (Carnivora, Mammalia) du Villafranchian d'Ain Brimba, Tunisie. *Bulletin du Muséum National d'Histoire Naturelle* 9:97–119.

Pilgrim, G. E. 1931. *Catalogue of the Pontian Carnivora of Europe in the Department of Geology*. London: British Museum of Natural History.

——. 1932. The fossil Carnivora of India. *Palaeontologia Indica* 18:1–232.

Rawn-Schatzinger, V. 1992. The scimitar cat *Homotherium* serum cope. Illinois State Museum Reports of Investigations 47:1–80.

Riggs, E. S. 1934. A new marsupial saber-tooth from the Pliocene of Argentina and its relationships to other South American predacious marsupials. *Transactions of the American Philosophical Society* 26:1–45.

Schaub, S. 1925. Über die Osteologie von *Machaerodus cultridens* Cuvier. *Eclogae Geologicae Helvetiae* 19:255–266.

Simpson, G. G. 1941. Large Pleistocene felines of North America. *American Museum Novitates* 1136:1–27.

Sotnikova, M. V. 1992. A new species of *Machairodus* from the late Miocene Kalmakpai locality in eastern Kazakhstan (USSR). *Annales Zoologici Fennici* 28:361–369.

Teilhard de Chardin, P. and J. Piveteau. 1930. Les mammifères fossiles de Nihowan (Chine). *Annales de Paléontologie* 19:1–132.

Turner, A. 1987. *Megantereon cultridens* from Plio-Pleistocene age deposits in Africa and Eurasia, with comments on dispersal and the possibility of a New World origin (Mammalia, Felidae, Machairodontinae). *Journal of Paleontology* 61:1256–1268.

——. 1993. New fossil carnivore remains. In C. K. Brain, ed., *Swartkrans: A Cave's Chronicle of Early Man*, pp. 151–165. Pretoria: Transvaal Museum Monograph No. 8.

Van Valkenburgh, B., F. Grady, and B. Kurtén. 1990. The Plio-Pleistocene cheetah-like cat *Miracinonyx inexpectatus* of North America. *Journal of Vertebrate Paleontology* 10:434–454.

MODERN SPECIES

Caro, T. M. 1994. *Cheetahs of the Serengeti Plains*. Chicago: University of Chicago Press.

Dallet, R. 1992. *Les Félins*. Paris: Nathan.

Dunstone, N. and L. Gorman, eds. 1993. *Mammals as Predators*. Oxford: Clarendon Press.

Eaton, R. L. 1974. *The Cheetah*. New York: Van Nostrand Reinhold.

Ewer, R. F. 1973. *The Carnivores*. London: Weidenfeld and Nicholson.

Gittleman, J. L., ed. 1989. *Carnivore Behaviour, Ecology and Evolution*. London: Chapman and Hall.

Guggisberg, C. 1975. *Wild Cats of the World*. New York: Taplinger Press.

Hemmer, H. 1978. Fossil history of living Felidae. *Carnivore* 2:58–61.

Herrington, S. J. 1986. Phylogenetic relationships of the wild cats of the World. Ph.D. diss., University of Kansas.

Hes, L. 1991. *The Leopards of Londolozi*. London: New Holland.

Hornocker, M. 1970. An analysis of mountain lion predation upon mule deer and elk in the Idaho Primitive Area. *Wildlife Monograph* 21:1–39.

Joubert, D. 1994. Lions of darkness. *National Geographic Magazine* 186 (2):35–53.

Kitchener, A. 1991. *The Natural History of the Wild Cats*. New York: Cornell University Press.

Kruuk, H. 1972. *The Spotted Hyena*. Chicago: University of Chicago Press.

Kruuk, H. and M. Turner. 1967. Comparative notes on predation by lion, leopard, cheetah and wild dog in the Serengeti area, East Africa. *Mammalia* 31:1–27.

Kurtén, B. 1973. Geographic variation in size in the puma (*Felis concolor*). *Commentationes Biologicae, Societas Scientarum Fennica* 63:1–8.

Leyhausen, P. 1979. *Cat Behavior*. New York: Garland STPM Press.

Mills, M. G. L. and Biggs, H. C. 1988. Prey apportionment and related ecological relationships between large carnivores in Kruger National Park. *Zoological Society Symposium* 65:253–268.

Neff, N. 1986. *The Big Cats*. New York: Abrams.

Pienaar, U. de V. 1969. Predator-prey relationships amongst the larger mammals of the Kruger National Park. *Koedoe* 12:108–176.

Rabinowitz, A. 1986. *Jaguar*. New York: Arbor House.

Rabinowitz, A. R. and B. G. Nottingham. 1986. Ecology and behaviour of the jaguar (*Panthera onca*) in Belize, Central America. *Journal of Zoology* 210:149–159.

Rudnai, J. A. 1973. *The Social Life of the Lion*. Lancaster: MTP Publishing.

Schaller, G. B. 1967. *The Deer and the Tiger*. Chicago: University of Chicago Press.

——. 1972. *The Serengeti Lion*. Chicago: University of Chicago Press.

——. 1993. *Tiger Crisis*. Bristol: BBC Films.

Seidensticker, J., M. Hornocker, W. Wiles, and J. Messick. 1973. Mountain lion social organization in the Idaho Primitive Area. *Wildlife Monograph* 35:1–60.

Seidensticker, J. and S. Lumpkin, eds. 1991. *Great Cats*. London: Merehurst.

Sunquist, M. E. 1981. The social organization of tigers (*Panthera tigris*) in Royal Chitawan Park, Nepal. *Smithsonian Contributions to Zoology* 336:1–98.

Thapar, V. 1986. *Tiger, Portrait of a Predator*. London: Collins.

——. 1989. *Tigers, The Secret Life*. London: Elm Tree Books.

Tilson, R. L. and U. S. Seal, eds. 1987. *Tigers of the World*. Park Ridge, N.J.: Noyes.

PALEOECOLOGY

Guthrie, D. 1990. *Frozen Fauna of the Mammoth Steppe: the Story of Blue Babe*. Chicago: University of Chicago Press.

Solounias, N. and B. Dawson-Saunders. 1988. Dietary adaptations and palaeoecology of the late Miocene ruminants from Pikermi and Samos in Greece. *Palaeogeography, Palaeoclimatology, Palaeoecology* 65:149–172.

Van Valkenburgh, B., M. F. Teaford, and A. Walker. 1990. Molar microwear and diet in large carnivores: Inferences concerning diet in the sabre-toothed cat, *Smilodon fatalis*. *Journal of Zoology* 222:319–340.

RECONSTRUCTION

Barone, R. 1967. La myologie du lion (*Panthera leo*). *Mammalia* 31:459–514.

——. 1986. *Anatomie Comparée des Mammifères Domestiques*. Vol. 1, *Ostéologie*. Paris: Vigot.

——. 1989. *Anatomie Comparée des Mammifères Domestiques.* Vol. 2, *Arthrologie et Myologie.* Paris: Vigot.

Bryant, H. N. and A. P. Russell. 1992. The role of phylogenetic analysis in the inference of unpreserved attributes of extinct taxa. *Philosophical Transactions of the Royal Society of London* B337:405–418.

Bryant, H. N. and K. L. Seymour. 1990. Observations and comments on the reliability of muscle reconstruction in fossil vertebrates. *Journal of Morphology* 206:109–117.

Ellenberger, W., H. Dittrich, and H. Baum. 1956. *An Atlas of Animal Anatomy for Artists.* New York: Dover.

Hildebrand, M. 1988. *Analysis of Vertebrate Structure.* 3d ed. New York: Wiley.

Knight, C. R. 1947. *Animal Drawing.* New York: Dover.

Muybridge, E. 1985. *Horses and Other Animals in Motion.* New York: Dover.

Spoor, C. F. 1985. Body proportions in Hyaenidae. *Anatomischer Anzeiger* 160:215–220.

Spoor, C. F. and D. M. Badoux. 1986. Descriptive and functional myology of the neck and forelimb of the striped hyena (*Hyaena hyaena*, L. 1758). *Anatomischer Anzeiger* 161:375–387.

——. 1988. Descriptive and functional myology of the back and hindlimb of the striped hyena (*Hyaena hyaena*, L. 1758). *Anatomischer Anzeiger* 167:313–321.

——. 1989. Descriptive and functional morphology of the locomotory apparatus of the spotted hyena (*Crocuta crocuta*, Erxleben 1977). *Anatomischer Anzeiger* 168:261–266.

Taxonomy

Berta, A. and H. Galiano. 1983. *Megantereon hesperus* from the late Hemphilian of Florida with remarks on the phylogenetic relationships of machairodonts (Mammalia, Felidae, Machairodontinae). *Journal of Paleontology* 57:892–899.

Bryant, H. N. 1991. Phylogenetic relationships and systematics of the Nimravidae (Carnivora). *Journal of Mammalogy* 72:56–78.

Cope, E. D. 1880. On the extinct cats of America. *American Naturalist* 14:833–858.

Flynn, J. and H. Galiano. 1982. Phylogeny of early Tertiary carnivora, with a description of a new species of *Protictis* from the Middle Eocene of northwestern Wyoming. *American Museum Novitates* 2725:1–64.

Hemmer, H. 1978. The evolutionary systematics of living Felidae: Present status and current problems. *Carnivore* 1:71–79.

Matthew, W. D. 1910. The phylogeny of the Felidae. *Bulletin of the American Museum of Natural History* 28:289–316.

O'Brien, S. J., G. E. Collier, R. E. Benveniste, W. G. Nash, A. K. Newman, J. M. Simonson, M. A. Eichelberger, U. S. Seal, D. Janssen, M. Bush, and D. E. Wildt. 1987. Setting the molecular clock in Felidae: The great cats, *Panthera.* In R. L. Tilson and U. S. Seal, eds., *Tigers of the World,* pp. 10–27. Park Ridge, N.J.: Noyes.

Peters, G. and M. H. Hast. 1994. Hyoid structure, laryngeal anatomy, and vocalization in felids (Mammalia: Carnivora: Felidae). *Zeitschrift für Säugetierkunde* 59:87–104.

INDEX

Page numbers for illustrations are in italics. All species are indexed by their scientific names.